Chilika

The Fishermen, the Catch, and the Challenges

Chilika
The Fishermen, the Catch, and the Challenges

Soumen Ray

AAKAR

Chilika: The Fishermen, the Catch, and the Challenges
Soumen Ray

First Published 2020

ISBN 978-93-5002-623-6

Published by
AAKAR BOOKS
28 E Pocket IV, Mayur Vihar Phase I
Delhi 110 091 India
www.aakarbooks.com

Laser Typeset at
Sakshi Computers, Delhi

Printed at
Sapra Brothers, Noida

Contents

Contents

Acknowledgments

The indescribable stories of fishers in Chilika that I had captured during my fieldwork have slowly and organically grown into this book over time. I was glad to finish with the exigencies of a PhD thesis, and personal commitments, before I set out to do some justice to Chilika Lake and its community of traditional fishers. The families I met, their wealth of knowledge, warmth, and care, so generously extended during my field study, remain my biggest gifts. It would be a fitting act of acknowledgement, if this book can create a more responsible and sustainable fishery eco-system in Chilika. While the fishers may disagree with some of my arguments, I am sure, they would take deep pride in the larger narrative.

This work would not have been possible without the guidance, support and encouragement of my guide, Dr. Rabindra Kumar Mohanty, Professor of Sociology in Mizoram University. With the responsibilities that my job entailed, I could hardly reach him in person for guidance; but he never let this distance be a barrier in the progress. His detailed study of my drafts, minute observations, valuable advice, constructive criticism, extensive discussions and interesting suggestions, have enhanced my knowledge immensely and kept me motivated. A special thanks to Dr. Amiya Kumar Das, Department of Sociology from Tezpur University for the support in sharing his experience and learning which helped me plan my work more meticulously. Dr. Das has generously shared his time and knowledge and played a major role in making me understand the art of structuring a book. I am also thankful to Dr. Rajshree Bedamatta, Associate Professor, IIT, Guwahati, who has given me her insights into capturing, formulating and articulating my analytical work and presenting concepts in a definitive manner.

Special thanks to officials of Chilika Development Authority (CDA), Department of Fishery and Animal Husbandry, NGOs, local administrative authorities of Chilika for their support during the research. Furthermore, I would also like to acknowledge with much appreciation the crucial role of administrative and technical staff members of the CDA, who were kind enough to advise and help in their respective roles. A special note of gratitude to Asish Sahu from Utkal University, Odisha, for making life fun while working. I would like to express my special thanks to him for being a great soul. He was always ready to help with a smile, especially during the data collection period.

I would also like to acknowledge the academic and technical support of the Department of Sociology, Utkal University. The library facilities and computer facilities of the University have been indispensable during my research. My special gratitude to Prof. Navaneeta Rath, Head, Department of Sociology, Prof. Anup Dash and Prof. D.N. Jena for their personal and academic guidance throughout the research. Special thanks to Dr. Gaurangi Maitra, (at gaurangimaitra.com), for reading, commenting, finalizing and editing the entire book. Any mistakes that remain however are mine alone.

I should also record my boundless gratitude to numerous friends and family in addition to the individuals already mentioned. With some I had the opportunity to share enthusiasm and curiosity, some provided inspiration, moral support and unconditional optimism, while others kept inquiring about the progress of the research and its practical purpose. I must thank all my colleagues, who have constantly encouraged me to finish this work and lent their support throughout with intellectual discussions. I must also thank UNFAO for bringing in wealth of knowledge in the fishery sector, which helped me in building my knowledge base.

Last, but not least, I would like to thank my wife and son, for their love, patience, and understanding. The book is dedicated to my parents for whom their son's journey might have been a source of pride and much amusement.

Prologue

This book began a decade ago, when I was selected for the ProVention Consortium Grants, an International Research Grants supported by the Asian Disaster Preparedness Centre (ADPC). I was excited by the grant design which engaged young professionals to form innovative links between research and action in disaster risk reduction. The project objectives certainly whet my appetite with opportunities and challenges that nurtured my core competence in disaster risk reduction in the development sector.

On a personal note, Chilika is an indelible part of our identity, lives, lore, legends, history, and aquatic ecosystems in Odisha. My earliest memories of Chilika, its natural beauty, and bounty came from the poems we recited as children. The lines from *Chilika Kavya* that described the fishermen coming back in the evening, rowing their boats to the rhythm of popular songs, left a lasting impression on my mind. As I write this prologue, I can vividly recall verses from this poem written by Kabibara Radhanath Ray (1848-1907), depicting water birds that are the special charm of the lake; resonating with the bells of cattle returning from the hills at dusk, and reflecting the moonlit hills and forests on the lake waters. These lyrical ballads brought alive the unique topography of Chilika long before I was able to visit in person and realize, "as a poet of nature Radhanath has done for Orissa, what Kalidas has done for India, putting the geography and the topography of his homeland to splendid poetic use" (Mayadhar Mansingh, 1962). When I first visited Chilika, I travelled by train and the journey brought back memories of the poem "Rela Upare Chilika Darsana" (Sighting of Chilika from a Train) by Utkalmani Pandit Gopabandhu Das (1877-1928). This multifaceted personality was simultaneously enthralled by the beauty of Chilika and saddened by his inability to see Chilika; he thus eloquently wrote of Chilika being the only place where the three eternal feelings of happiness, solace, and love could be found.

It is a matter of pride that Chilika is the largest coastal lake in India and the second largest brackish water lake in the world (Inventory of Wetlands, 2018). Akin to other global water bodies, Chilika has borne witness to primary small-scale household enterprises surviving beside large-scale commercial sectors in fisheries and aquaculture. Chilika is therefore a significant part of the national open water resources (excluding the oceans) and contributes to food security, poverty alleviation and social well-being of millions of fishers in the Indian subcontinent. Yet, the impact of mechanization, the emerging threats to common property with open access, and environmental hazards, in the context of alternate livelihood options have received inadequate attention from social scientists. As a direct causatum of these deficiencies, the objectives of my investigation explored the complexities in the context of property rights; and examined alternative sustainable livelihood options for the fishers in the Chilika region.

At the very core of my research were personal interactions with fishers and their families, which remain my most vivid recollections of the Chilika region. Our preliminary conversations on where and when the fish would breed, migrate, and regenerate, became good indicators of the depth of their indigenous knowledge. They held a wealth of information about the past, present and future ecology of the community; with different age groups subscribing to different value systems and indigenous beliefs regarding fishery management. As the picture of their understanding of fishery resources emerged, I realized they could contribute substantially to the contemporary discussions around sustainable fishery management. Quite often, our discussions on the heightened vulnerability of small-scale traditional fishers would converge on climate induced natural disasters and fishery livelihoods; and on the entry of big business houses with large-scale mechanization and deep pockets. On several occasions, they were not only highlighting their subjection in dealing with the current situations, but a few of the fishers became increasingly agitated when the discussion touched on property rights. Not unnaturally, I became eager to undertake an in-depth study, which began with detailed fieldwork during 2009-10, for my PhD thesis.

It became increasingly clear to me that it is vital to comprehend and objectively evaluate the fishers' thoughts and perspectives, their extant resources, and the complexities involved. Subsequently this knowledge could become an integral part of their vulnerability reduction from the external environment and climatic variation. There were several intrinsic issues, which I could not highlight in my thesis, which in turn led me to writing this book. Yet, I am more than aware that no one book can do justice to the wealth of the fishers' knowledge that exists, and their ever increasing livelihood vulnerabilities due to increased competition, unregulated property rights and climate change.

REFERENCES

1. Mansingh Mayadhar, *History of Oriya Literature*. New Delhi: Sahitya Academy, 1962, p. 185.
2. "Inventory of Wetlands", Government of India, pp. 314-318. Accessed on July 15, 2018, http://envfor.nic.in/divisions/csurv/Wetland Inventory.pdf

1

Casting the Net in Global and Regional Waters

As children, we waited eagerly for the festivities that marked Kartik Purnima (the full moon in the months of November–December); especially, the happy ritual of making miniature boats and racing them across the nearest body of water. Our elders would be more sacrosanct, making boats (*boita*) out of banana stems and coconut sticks, filling them with earthen lamps, fabric, and betel leaves, as prescribed for the ritual of *Boita Bandana.* The boats would then be set afloat with lamps lit and prayers; these benedictions animated the waters and even quietened us as restless youngsters for a while, before we went back to racing our boats again. It was only very much later, I read Sanjeev Sanyal's book, *The Ocean of Churn: How the Indian Ocean Shaped Human History* (2016). It made me realize that the annual *Boita Bandana* recreated benediction rituals meant for sailors leaving on long voyages that began in Chilika in the second century BC. The southern route to the Indian Ocean began at Chilika, where sailors from India were able to take advantage of the monsoon winds and ocean currents to trade with Indonesia, Java, Sumatra and Bali.

This was just one among the many stories, poems, and songs, not to mention its aquatic bounty, that made Chilika part of our common vocabulary in Odisha. Thus it was only natural that Chilika formed the core of my project on, "The indigenous fishery livelihood in Chilika Lake: and the development challenges in its resource management." Given the broad-based context of Chilika, an overview of the fisheries sector in the global and regional contexts, with the later devolving sequentially on India, Odisha, and Chilika, would be germane to casting our nets before we began fishing in earnest.

Globally millions of people depend on fisheries, aquaculture, and the related tertiary sectors as important means of earning and building livelihoods. Fishery-related livelihoods by their unregulated nature are multifaceted, self-motivated, and adaptive; given their dependence on open resources, including oceans, rivers, lakes and similar common property. Estimates of the global population employed directly and indirectly this sector vary from 140 to 260 million persons (FAO, Teh and Sumaila, 2013 respectively), with the latter including the small-scale fishers. Irrespective of these contradictory numbers, the fishery sector is undoubtedly a vital segment in creating livelihood opportunities for people in many parts of the world. Despite their significant contribution to the world fishery sector, small-scale fishing communities are often impoverished by rampant over fishing and critical factors like insecure access rights to fishery resources; poor or absent health and educational services; lack of social safety nets; vulnerability to natural disasters and climate change; and exclusion from wider development processes (FAO, 2012). Therefore, due diligence to fishery regulation and socio-economic implications are cardinal to sustainability in global fishery communities. And yet, this is only half the story.

The other half comes from the major shift within the fishery sector from a 'production growth approach' to 'improved fisheries management', which began in the mid-1900s. The focus during this period was to address fishing pressure, using various regulations, so that sustainable catch levels could be achieved in perpetuity. Escalation in fishing was initially a post Second World War phenomenon, partly fuelled by highly proficient fishing methods and technology, even when most of it came from coastal waters (Sumaila et al. 2013). Increased demand for fish, growing volume of business, and subsidized expansion of fishing capacity, spurred a manifold increase in the fisheries sector. The consensus that humans could never overfish has apparently been proved wrong along with marginalization of resource groups and impending ecological impact. Consequently, the management and governance systems for fisheries aimed specifically at state-regulated fishery, evolved further in the twentieth century, empowered by mutual understanding and consensus. The document from the UN Session on Sustainable Development, entitled, "The Future We Want", underlines the vital role of small-scale fisheries for sustainable development and highlights the need for inclusive empowerment.

In 2015, fish accounted for about 17 per cent of the global population's intake of animal protein and 7 per cent of all protein consumed. In addition, it provided more than 3.2 billion people with almost 20 per cent of their average per capita intake of animal protein (*State of World Fisheries and Aquaculture* (SOWFA, 2018). However, average per capita fish intake may vary significantly within regions, based on the culture, practices, economy, and key geographic factors[1]. For example, across countries, annual per capita fish consumption varies from less than 1 kg to more than 100 kg (SOWFA, 2018). Per capita apparent consumption for world fisheries and aquaculture, as calculated by FAO is shown in Figure 1.1.

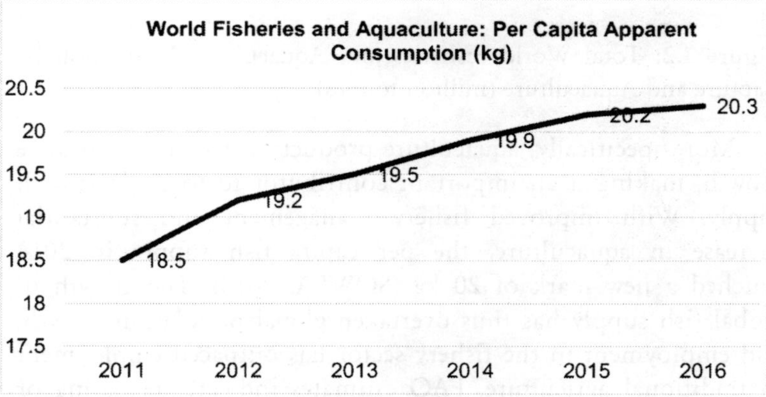

Figure 1.1: World Fisheries and Aquaculture: Per Capita Apparent Consumption (kg) Source: *The State of World Fisheries and Aquaculture*, 2018, FAO.

As per FAO (SOWFA, 2018) data, (Figure 1.2) the global fish production* peaked at about 171 million tonnes in 2016, with aquaculture representing 47 per cent and capture representing 53 per cent of the total. During this period, while the production through capture method remained comparatively static, aquaculture displayed a constant increase (almost 5.8 per cent annual growth between 2011 and 2016), and has been responsible for the continued growth in the supply of fish. The agency has further estimated that the production is expected to grow by 18 per cent to 201 million tonnes by 2030.

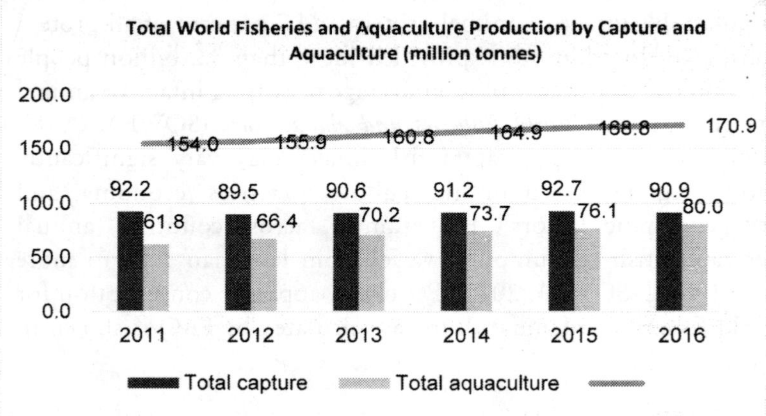

Figure 1.2: Total World Fisheries and Aquaculture Production by Capture and Aquaculture (million tonnes).

More specifically, aquaculture production has seen a massive growth, making it an important contributor to total global fish supply. With improved fishery management and substantial increase in aquaculture, the per capita fish supply in 2014 touched a new mark of 20 kg (SOWFA, 2016). The growth in global fish supply has thus overtaken global population growth; and employment in the fishery sector has outpaced employment in traditional agriculture. FAO estimates indicate stagnating or decreasing share of employment in capture fisheries compared to the aquaculture sector, where increased employment opportunities are available (SOWFA, 2016).

The fisheries as a sector are managed by several stakeholders including local, national, regional, and global organizations with varied interests that depend on the nature of their professional domain. Hence, different levels of control mechanisms, starting from international organizations to local fishery cooperatives have been raised and strengthened to manage fisheries and fishery-related livelihoods. Collectively, the governance of fisheries has achieved significant progress through national and international efforts with the objective of sustaining the sector. At the global level, the United Nations Convention on the Law of the Sea (UNCLOS 1994), is the fundamental regulator of international fisheries management. It provides a legal framework for the world's fishery sector, stipulates rights and accountability of marine countries towards resources sustainability; while protecting and preserving the marine environment (UN, 2010). In addition, the UN Fish Stocks Agreement, 2002, mandated the international community to ensure long-term conservation and sustainable use of fishery stocks. During last few decades, international communities have taken up regulation of the world fishery resources to counter climate change, overfishing, and pollution; which are endangering recent gains in protecting the world's large water bodies and ultimately impacting global fishing. For example, the International Conference on Nutrition held during 2014, while strongly highlighting the value of fish and seafood as an important source of proteins and essential micronutrients for women and young children, also endorsed greater management and regulation of fishing practices to safeguard nutritive value and maintain healthy intakes.

Fisheries in India: A Historical Overview

The wealth of freshwater and marine resources of India that ensure the fisheries sector is a significant contributor to the livelihood of millions and its economy. The British era Indian Fisheries Act of 1897 was an early pan Indian legislation which made fisheries a state subject, but minimal progress ensued. The erstwhile Madras Presidency (Mosse, 1998) and Bengal Presidency showed individual levels of progress in small-scale fisheries and fresh water aquaculture, respectively. The linking of agrarian policies with the management of water and potential for fisheries of this past era, finds a continuum in the Indian Council for Agricultural Research (ICAR); with the latter still holding the policy making authority for the fisheries sector in independent India.

Post-1947, a series of intensive efforts that included organized research and development, were undertaken; with emphasis on the Indian fishery sector growth through modernization and intensive capital investment (Kurien, 1978). Subsequently, the Central Fisheries Research Institutes for Marine and Inland Fisheries, the Deep Sea Fishing Station, the Central Institute of Fisheries Technology, and the Directorate of Cold-Water Fisheries were established. An analysis of the Government of India's Five Year Plans endorsed the urgency for fishery development and provided specific recommendations, which consequently resulted in the Indian Fishery sector posting growth induced by reforms.

The first three Five Year Plans focused mostly on the marine fisheries sector with specific emphasis on the mechanization of native crafts, introduction of automated fishing boats, modernization of fishing equipment and adoption of synthetic tools for augmented fishing. Concurrently, infrastructure for handling plants, cold storages, landing, processing, transportation, and marketing facilities were created. These concerted measures, combined with intensive scientific studies, documentation and strengthening of various other facets, set the marine fishery sector on a growth trajectory. Simultaneously, the global impact of improved communication and the increased demand for fish, championed the fishery market in India. During the Fourth Five Year Plan period (1969-74), India initiated deep-sea fishing through introduction of trawlers. At the same time the inland fishery further received a boost through positive and successful development of improved breeding techniques.

In the inland fishery sector, several factors like the establishment of Fish Farmers Development Agencies, and the creation of seed production centres, particularly enhanced both culture and freshwater fishery. The introduction of technology spawned supplementary industrial units for boat, net, and diesel engine production for a burgeoning fishery sector. Supported by the increased volumes of export, the fisheries sector in India showed substantial growth, vindicating the developmental measures undertaken during 1947-77 by the Government of India. Subsequently, the fishery sector helped the country to build up a fishing industry, which played a key role in the overall growth of the country.

Going beyond increased production targets, the Fifth Five Year Plan (1974-78) endorsed the importance of fish products for meeting nutritional requirements at the household level. This in turn resulted in higher economic growth, improvement in the lives of fishermen, and increased foreign exchange income from exporting fish and fishery-related products. This plan stressed cohesive and balanced development of marine fishery to maximize return on investment. In 1976, the provision for infrastructural facilities was intensified with the announcement of an Exclusive Economic Zone for deep-sea fishing. During this period several other developmental strategies were undertaken including the expansion of fishing to the areas which were otherwise lying unused, augmentation of fishing and fishery products and improvement of traditional fisheries.

In the late 1980s and early 90s, two separate institutions, i.e. the Brackish Water Fishery Development Authority (BFDA) and Marine Products Export Development Authority (MPEDA) were established in India to further promote and effectively manage shrimp culture. In addition, the 1986 Environment Protection Act of India put in place conservational measures to prevent impoverishment of the fishery resources. With the increased focus on aquaculture, the Government of India was able to mobilize additional support from multinational organizations and several developed countries, including Britain. Therefore, by the mid-90s, shrimp mono culture was enhanced and simultaneously commercial culture of brackish water shrimp gained momentum in India. The emphasis on shrimp aquaculture and the resultant resources generated, made Martinez-Alier claim that the shrimp aquaculture resulted in a situation where the 'Blue Revolution' produced 'Pink Gold' (Martinez-Alier, 2002).

Yet, the story of shrimp cultivation in India dates back to the 1960s as detailed by Kurien (1978); when after an initial lag phase, production rose with intensive mechanization and support from the Norwegians, along the coast of Kerala. The Norwegians had envisioned the export potential of dried shrimp to Sri Lanka and Burma and fresh shrimps to European markets. This necessitated further mechanization of the fishing vessels, establishment of freezing plants, and trade agreements with different countries for shrimp export. Kurien argues that between 1960 and 1974 shrimps jumped from 30 per cent to 77 per cent of the total marine export of India; and in terms of value, they rocketed from 41 per cent to 90 per cent. The unit value realized increased from Rs. 3.32 kg to Rs. 19.10 kg during the same period. For traditional fishermen using non-mechanized crafts, though shrimps were a regular catch, it was hardly ever consumed, but used instead as manure for coconut plantations. Therefore, the sudden surge in demand for shrimps was a revelation in terms of ready cash. They now had the capacity to increase their catch, and avail of the liberal off season cash advances, for a mere pledge to sell their shrimp catch to the same merchant or middleman. Till the late 1960s this class of producers contributed over three-quarters of the shrimp catch. Thus, the seemingly insignificant shrimp linked the traditional fishers to the world market through a hierarchy of middlemen in India, culminating with the rich merchant capitalist owners of freezing plants and insulated vans. The frozen shrimps travelled in the latter to agencies abroad, passed through chains of department stores and frozen food retail outlets, to the final ultra-modern consumers in the developed industrial countries. The increasing export and demand for shrimps endorsed the promotional role given to the shrimp industry by the Government of India. By 1997-98, India earned Rs. 4,120 crores as income out of fishery exports of which more than 50 per cent was solely contributed by export of shrimps. In 2006, under the Ministry of Agriculture, Government of India initiated a dedicated organization, focused solely on fisheries.

The approach paper of the Twelfth Five Year Plan highlighted the success of the fishery sector and the doubling of fish prices in the

Indian market in the previous Eleventh Plan period. Remarkably, the price rise in the fishery sector was much higher compared to either crops or any other livestock. The approach paper further emphasized that the major growth in fisheries in the Eleventh Five Year Plan had come from the inland fisheries; with the burgeoning shrimp export potential spawning swift development of brackish water aquaculture. Over time, the government's Five Year Plans and their implementation gradually shifted from producing greater quantities of fish to improved culturing and responsible fishery. Introduction of modern fishing technology supported by evidenced based policies and practices, significantly improved the socio-economic conditions of the traditional fishers. Against this backdrop, it is important to locate the development of fisheries within the state of Odisha.

Odisha: An Overview of the Fishery Sector

Odisha, a major maritime state, within which Chilika is located, has a 480 km long coastline, which provides enormous scope for development of inland water and marine fisheries. The state has six maritime districts (Balasore, Bhadrak, Kendrapara, Jagatsinghpur, Puri and Ganjam) with Puri district covering more than a third of the coastline (Department of Fishery (DOF), Government of Odisha, 1998). Figure 1.3 Odisha Map.

The long coastline can be divided into two distinct areas: the northern and southern coasts. The former is a shallow coast extending northwards from Kendrapara district to Balasore district. This area has a broad shelf, gradual slope and greater tidal activity; with muddy calm waters, characteristic of its tidal flats and extensive river deltas. The latter is the southern coast extending southwards from Jagatsinghpur district to Ganjam district, which is narrower, with broad sandy beaches and open surf-beaten shores (FAO Bay of Bengal Programme (BOBP), 1984a; 1986a; Ayyappan and Jena, 2000).

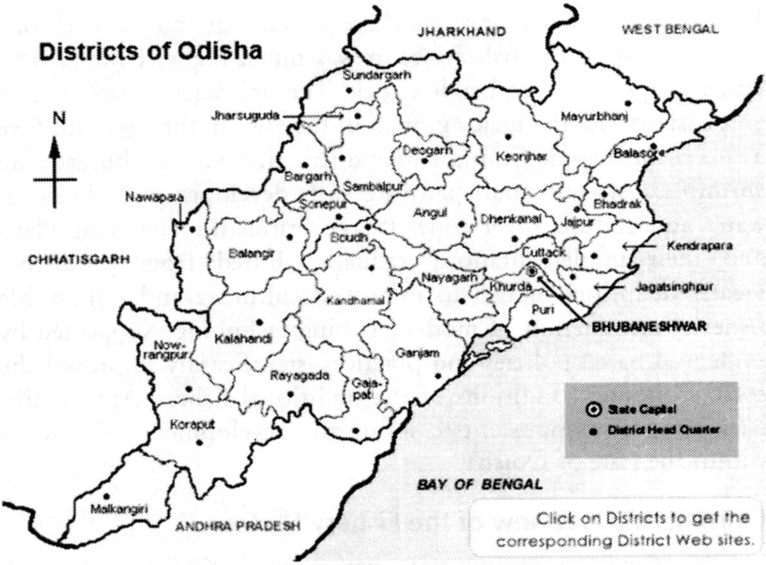

Figure 1.3: Odisha State Map with Districts, Source: Government of Odisha (https://www.odisha.gov.in/content/dist), accessed on March 22, 2020.

The Government of Odisha fishery policy, 2015, states that the freshwater resources of the state include about 6.76 lakh ha area under tanks and ponds, lakes and reservoirs, swamps and jheels, besides a network of rivers and canals. In addition, there exists 4.18 lakh ha brackish water resources, which are divided between Chilika Lake (0.79 lakh ha), estuaries (2.98 lakh ha), brackish water areas (32,587 ha) and the backwaters (8,100 ha) (Odisha Fishery Policy, 2015). By use of these resources, the fish production from capture and culture-based fisheries have sustained livelihoods of millions of fishers in the state and substantially augmented domestic demand. During the past several years the state has also exported substantive volumes of fishery products to global markets, consequently leading to increase in employment and income-generating opportunities for the rural poor.

According to the *Handbook on Fisheries Statistics of Odisha, 2000/01* (DOF, 2002), Odisha has a total of 589 marine and 3,289 inland fishing villages; with the number of fishers in the state touching over 1 million. The report also highlights that there are about 86,000 active marine fishers in the state. These figures are somewhat contradictory, because in the two previous year books of DOF, for 1992/93 and 1996/97, the number of marine fishing villages was estimated to be 329 and the inland fishing villages were 6,895 and 6,899 respectively. It is possible that in the post-1999 cyclone period, many inland fishing villages were included in the marine category to secure eligibility for support under various relief and rehabilitation packages. This could also partly explain the decline in the number of inland fishing villages during the period. However, the *Economic Survey of Odisha, 2011-12* shows there are 3,678 fishing villages in the state, out of which 132 villages are in Chilika.

The total fish production of the state in 2013-14 was 4,13,000 tonnes. The overall increase in fish production during the five-year period from 2007-08 to 2012-13 was in the order of 63,660 tonnes, with an annual growth rate of about 3 per cent. While the fish production from the inland fisheries (freshwater and brackish water) resources has increased from 2,18,716 tonnes to 2,94,000 tonnes, the marine fish production declined from 1,30,767 tonnes to 1,20,000 tonnes during the corresponding period. The annual per capita fish consumption presently is 9.13 kg in Odisha as against the national per capita fish of 9.8 kg. The brackish water resources of the state are confined to 7 coastal districts (including Khurda district). Out of the total potential area of 38575 ha, only 32587 ha, is considered suitable for brackish water aquaculture.

Fisheries of Odisha around Chilika, were under the Director of Industry, Government of Odisha from 1941 to 1953, but in 1956 a separate Directorate of Fisheries was formed. The plans, programmes and policies of the fishery sector have subsequently been carried out through the Directorate of Fisheries, Odisha. The Directorate of Fisheries is a service sector oriented technical department, promoting scientific aquaculture in the state and the welfare of fishers. Over time, with the increase of several tertiary sector activities and support for the growth of the fishery sector; the rationale underlying fisheries management, their exploitation and development are beginning to change, presenting new challenges to the fisheries administration. Vital operational regulatory instruments, such as the Reservoir Fisheries Policy and the Orissa Marine Fishing Regulation Act (OMFRA), do not consider some important aspects of sustainable fisheries development and management in the contemporary context.

Fisheries, in the state of Odisha contribute to multiple livelihood options; directly as food, as a source of income, and through other social benefits, such as reduced vulnerability to poverty. However, both the lake and marine fishers are analogous as they are rendered non-operational without the open access nature of resources. Small-scale fishers in Odisha are characterized as poorest of the poor and the most vulnerable; and management of open resources by the state and mechanization have further contributed to their marginalization, and adversely impacted the traditional and small-scale fisheries. In addition, the entry of landless agriculture labourers (without alternative livelihood options) into the fishery sector has intensified the competition for limited resources. This underplays the extent to which structural biases systematically produce vulnerability (Hickey and du Toit, 2006). Further, the increased exposure to environmental shocks and hazards makes the fishers more vulnerable particularly because of lack of alternate livelihood options available. For example, it is expected that the recent increase in high seasonal temperatures caused by climate change can dramatically impact fishery livelihood in the coastal areas across the world, and the fishers in Odisha.

The fishery sector in Odisha is undermined by high levels of inequality of access between traditional fishers and private farms. The entry of large private farms poses huge challenges to traditional fishers, who have an inherent potential to generate more catch sustainably. Thus, in terms of securing loans and other services, private and large businesses are benefited due to economies of scale.

In addition, the socio-economic and cultural aspects of several traditional fishers are under risk on account of liberalization in the fishery sector. It is often claimed that concerns and opinions of small fishers have not been given serious consideration in framing policies and guidelines. The challenge in this context is to develop a long-term policy and strategy that incorporates balanced and sustainable management, and development of the fishery sector in Odisha.

With this background, it is now possible for us to look closely at the region which principally constitutes the study area of my book. Humanitarian, climatic and livelihood security concerns related to the fishery sector in Odisha demand a commitment for dedicated focus and prompt action for building resilience to climate shocks, preparedness for adaptation and production volatility. A progressive and new policy environment could spur overall development of the fisheries sector.

Chilika Lake, the Main Resource Arena

Chilika Lake is the largest brackish water lake in India and the second largest in the world, situated in the eastern part of Odisha on the Bay of Bengal. Surrounded by the three coastal districts of Puri, Khorda and Ganjam, Chilika is a highly climate sensitive area and considered the largest winter ground for migratory birds in the Indian subcontinent. In 1993, it was placed on the Montreux Record (Ramsar's list of sites undergoing ecological degradation); subsequently due to the positive restoration of the lake's ecosystem by the Chilika Development Authority, it was awarded the Ramsar Wetland Conservation Award in 2002 and removed from the Montreux Record. The lagoon has a long and rich history of sustainable fishing. The fishers in Chilika have been in the fishery business for several generations.

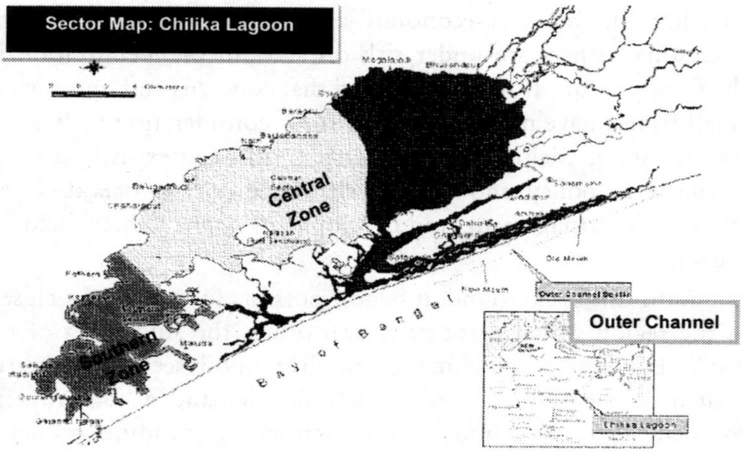

Figure 1.4: Sector Map of Chilika Lagoon, Source: Chilika Development Authority (https://www.chilika.com)

The lagoon can be broadly divided into four natural (ecological) zones based on salinity and depth; the southern zone, central zone, northern zone and the outer channel. Basically, fresh water dominates the northern zone, brackish water the western zone, while the southern zone is designated as a zone characterized by higher salinity. The outer channel is a marine zone in the non-monsoon periods; and during the monsoons, flood water flowing into the lagoon converts it into a freshwater zone. This flood water is finally released into the sea through the lagoon's mouth. The lagoon is also influenced by three sub-systems, the Mahanadi River system in the north, the 52 channels of the western catchment river systems, and the Bay of Bengal on the east. The Chilika drainage basin, including the lagoon itself, covers an area of over 4,300 km. (Das and Samal 1988). The watershed boundaries lie between the water flowing into the Mahanadi and Chilika in the north, while the areas draining into the Bhargavi River make up the north-east watershed; in the west and south-west, the watershed boundary lies between the streams flowing into the Rushikulya River, and those flowing into Chilika (Ram et al., 1994). The salinity level in the lagoon is thus dictated by river discharges during different seasons, wind action, and the extent of tidal influx carrying marine water from the Bay of Bengal. The seasonality of rainfall also leads to large fluctuations in the size of the lagoon.

The size of the lake fluctuates substantially within the course of a year, from a maximum of 1,165 km during the monsoon season, to a minimum of 906 km during the dry season. A channel of about 1.5 km width runs parallel to the Bay of Bengal and is separated from it by a narrow spit. A 32 km long, narrow, outer channel connects the lagoon to the Bay of Bengal. In 2000, the Chilika Development Authority created a new mouth with the Bay of Bengal at Satpada. Further dredging was also made in the channel between the outer channel and the rest of the lake for effective water passage and to maintain salinity. This decision resulted in the improvement of the lake's water quality in terms of salinity. After the opening of the new mouth in Chilika the fish production from brackish water showed a two-fold increase from 2001 onwards. Scholars such as Ghosh (2006) argue that re-salinization improved the ecosystem functioning. The opposing view states that salinization has led to a reduction of ecological diversity which might adversely impact livelihoods and their sustainability in the future. Such contrasting views are subject to greater scrutiny in a subsequent chapter of this book.

The Chilika lagoon with its unique combination of marine, brackish, fresh water and estuarine characteristics, is a principal source of capture fisheries of Odisha, which supports food and livelihood security of more than 1,00,000 fishers living in and around the lagoon. A direct use benefit of the Chilika ecosystem, amounts to its fisheries output being more than 71 per cent of its economic value (*Chilika Development Authority Report, 2009*). It provides employment opportunities to large numbers of local artisans and small-scale business communities in the support service sector. It also contributes a substantial share to Odisha's shrimp export.

It is equally important to mention that Chilika Lagoon is a climate sensitive area in India. Although limited quantitative data on the effects of climate change in Chilika Lagoon are available, floods, droughts, heat waves, and sometimes cyclones are common features in the state. The lagoon experiences south-west and north-east monsoons during June to September and November to December respectively. December to February is the winter season which is followed by the hot season from March to May. The period from June to September is the monsoon season while October and November months are the post monsoon transition months (Chilika Development Authority (CDA), 2010).

The lake is surrounded with approximately 132 fishing villages of varying sizes and approximately 30 per cent of their population is actively involved in fishing. Both fisher and non-fisher people living in and around Chilika Lake are involved in fishing activities. The fishermen of Chilika belong to different major sub-castes: (i) Khatia, (ii) Keuta, (iii) Kandara, (iv) Tiara, (v) Kartia, (vi) Nolia and (vii) Bengali refugees. Most of the fishermen belong to the Keuta and Kandara sub-castes; with the Keuta, Kandara and Tiara belonging to Scheduled Castes. The Kartias mainly operate bamboo screen traps, though some of them use the fishing nets too. The Nolias live between the sea coast and the banks of Chilika; they fish in both the sea and the lake and are Telugu-speaking fisher folk. Large refugee populations from East Bengal, who have settled on the western banks of Chilika since the 1950s have adopted fishing as a means of their livelihood. They use boats, nets and traps to capture fish based on their requirements. Important types of traps used locally are called 'Baza', 'Dhaudi', 'Jana', 'Dian', 'Uthapani', 'Bahani', etc. The traditional fisher people occupy the lowest position in the social strata; their poverty reinforcing their lower social position.

Besides fishing, there are some fishers who also use their boats for tourist activities in the lake. Motorized boats ply four to five times a day from Chamutia to Satpada via Mahisha or from Berhampur to Satpada; with the ghat controlled by the local Gram Panchayat. During our fieldwork from 2008-10, besides motor launch services from Balugaon to Kalijai on Sunday, one could reserve a motorized boat from Satpada to Balugaon via Nalabana and back or vice-verse with a payment of Rs. 800 to Rs. 1,000. A fee of Rs. 250 is charged by the dolphin association for one and a half hours to see dolphins near Satapada Jetty (Chilika Development

Authority, 2001). Similarly, there are several other places from where the motorized boats can be hired for sailing on the lake to watch migratory birds or visit different islands.

While the fish production in Chilika Lake has increased, the underlying principles for fisheries management, growth and development is changing. The state of Odisha is yet to direct its focus on responsible fishing and sustainable management; as the extant goals have been wealth and revenue generation and their distribution, which poses new challenges to the management of fisheries. As I have discussed earlier in this chapter, the development challenges have evolved, and the understanding of the concerns on sustainable development have been further strengthened. I take up the question of policy in a subsequent chapter of this book. In any case, one cannot venture to speak on policy issues without a more complete understanding of the environmental issues that pose obstacles to development of the region. In the subsequent chapter I have delved into some of the key environmental issues in the region, thus making it possible to make a socio-ecological assessment of the fishery livelihoods in Chilika. The traditional fisher folk and the Chilika Lake have been particularly hard hit by several socio-political, economic and environmental problems that were significant and warranted examination. Concurrently, a change in government policy led to the leasing of fishing grounds for commercial shrimp farming, which in turn disrupted the fishing rights and access to fishing grounds of the traditional fishers. Conflicts over this issue have resulted in violent clashes and loss of life over certain periods of time. On the socio-economic front, several studies have been conducted on the fishing community of Chilika mostly focusing on community life, business promotion, environmental sustainability and livelihood issues. However, the impact of mechanization and the emerging threats to common property with open access, the natural disaster and the environmental hazards in the context of alternate livelihood options have not attracted the attention of researchers to the same degree. Most of the research focused on the primary production phase of the fishing industry, has so far been in the field of biology.

Owing to the lack of theoretically substantiated economic research and limited analytical tools, biologists have extended the scope of their own thoughts and findings into the economic sphere and in some cases have penetrated deeply. Many others who have paid no attention to the economic aspects of the problem have nevertheless recognized that the ultimate question is not the ecology of life in the lake as such, but the human exploitation of Chilika's resources for their own economic gains.

Some of the researchers have given greater attention to the economics of fishing and have viewed the fisher people not as an irresolvable community but as a part of the peasant society and the wider market system. Others have concentrated on the technology of fishing without looking at the disruption it has brought to the life and livelihood of fisher people operating manual boats. Some studies have sought to portray the social life of fishermen at a specific point of time, while others have tried to unfold the pace, degree and direction of the change taking place in their social life over a certain span of time. Some of these are descriptive ethnography studies and only a few of them are post specific hypothesis to seek the ethnographic data in a theoretical framework. It is also evident that most of the studies on the lake are based on ecology, as the issues have drawn the attention of many of the eco-activists and the region has witnessed many forms of social protests and movements by the fishermen's community. It is also true that fisher people's livelihood is affected in the wake of commercial exploitation of resources in the lake. However, very few studies have been conducted about an alternative livelihood for fisher people. Similarly, no specific studies have been conducted on the collective movements and its relation to the concept of common property resources. Though quite a few studies have been conducted in the area of dynamics of access to fisheries resources in Chilika, they have failed to justify the co-relation between the various social movements and their relation to the right of fishers. The common focus area of almost all the studies are "issues such as conflicts between the young and old fishermen, the introduction of mechanized boats and the state versus the community management".

One of the most important areas, which have completely remained out of focus in any of the research, is natural disasters which have a major impact on the fisher people especially due to constant ecological degradation.

The first chapter has viewed the fisheries sector through its multiple stakeholders and control mechanisms, from international organizations to local fishery cooperatives in the context of fisheries and fishery-related livelihood management. An overview of the fisheries sector in the global and regional contexts, with the latter devolving first on India, then Odisha, and finally Chilika has been outlined. Topography, ecology, and the fishery potential of Chilika Lake and disruption of traditional livelihood through shrimp culture have been dealt with in detail. The major shift within the fishery sector from a 'production growth approach' to 'improved fisheries management' has also been examined especially against the significant contribution to the world fishery sector by small-scale fishing communities. Their socio-economic well-being through regulated fishing and access rights; provisions for healthcare, educational services, social safety nets, disaster management and climate risk mitigation; dangers of mechanization, shrimp farming and exclusion from wider development processes will be developed in the following chapters. In a final analysis of this opening chapter, it can be concluded that due diligence to fishery regulation and socio-economic implications are cardinal to sustainability in global fishery communities.

NOTE

1. Globally, fish consumption grew from 9.0 kg in 1961 to 20.2 kg in 2015, at an average rate of about 1.5 per cent per year (SOWFA, 2018).

REFERENCES

1. *The State of World Fisheries and Aquaculture (2016), Contributing to Food Security and Nutrition for All.* Rome: Food and Agriculture Organization of United Nations.

2. *The State of World Fisheries and Aquaculture (2018), Meeting the Sustainable Development Goals.* Rome: Food and Agriculture Organization of United Nations.
3. Sumaila, U.R., Lam, V., Le Manach, F., Swartz, W., and Pauly, D. (2013), "Global Fisheries Subsidies." European Parliament, http://www.europarl.europa.eu/RegData/etudes/note/join/2013/513978/IPOL-PECH_NT (2013) 513978_EN.pdf.
4. UN (2010), Fish Stocks Agreement: Overview of What the Agreement Says and its Impact; Published by the United Nations Department of Public Information | May 2010 | DPI/2556 B http://www.un.org/depts/los/convention_agreements/reviewconf/FishStocks_EN_B.pdf
5. Teh, L.C.L. and Sumaila, U.R. (2013), Contribution of Marine Fisheries to Worldwide Employment. Fish and Fisheries, 14: 77–88. doi:10.1111/j.1467-2979. 2011.00450.x
6. FAO (2012), Towards the Future We Want; End Hunger and Make the Transition to Sustainable Agricultural and Food Systems. Published by Chief, Publishing Policy and Support Branch Office of Knowledge Exchange, Research and Extension, FAO, Rome, Italy. http://www.fao.org/docrep/015/an894e/an894e00.pdf
7. Mosse, D. (1998), Making and Misconceiving Community in South Indian Tank Irrigation, Paper Presented at CROSSING BOUNDARIES, Conference of the International Association for the Study of Common Property, Vancouver, June 10-14, 1998 http://dlc.dlib.indiana.edu/dlc/bitstream/handle/10535/882/mosse.pdf?sequence=1
8. Kurien, J. (1978), Entry of Big Business into Fishing: Its Impact on Fish Economy. *Economic and Political Weekly*, September 9.
9. Martinez-Alier (2002), *The Environmentalism of the Poor.* Chektenham: Edward Elgar.
10. DOF (1998), *Handbook on Fisheries Statistics of Orissa, 1996/97.* Bhubaneswar: Directorate of Fisheries, Government of Orissa.
11. BOBP (1984a), Marine Small-Scale Fisheries of Orissa: A General Description. BOBP/INF/7.Madras.
12. BOBP (1986a), Traditional Marine Fishing Craft and Gear of Orissa, BOBP/WP/24.Madras.

13. Ayyappan, S. and Jena, J.K. (2000), Coastal Fisheries in Orissa, an Important Economic Activity. In V.K. Sharma, ed. *Environmental Problems of Coastal Areas in India*. New Delhi: Bookwell.
14. Odisha Fishery Policy, 2015, Department of Fisheries, Government of Odisha http://investodisha.gov.in/download/Odisha_Fisheries_Policy_2015.pdf
15. DOF. (2002), *Handbook on Fisheries Statistics of Orissa, 2000/01*. Bhubaneswar: Directorate of Fisheries, Government of Orissa.
16. Hickey and du Toit (2006), Adverse Incorporation, Social Exclusion and Chronic Poverty, CPRC Working Paper 81 http://www.chronicpoverty.org/uploads/publication_files/WP81_Hickey_duToit.pdf
17. Ghosh et al. (2006), Restoring Ecological Balance and Livelihoods Through Re-salinization. *Lakes & Reservoirs: Research & Management*, 11: 239–255. Published Online.
18. Das, N.K., and Samal, R.C. (1988), Environmental Survey of Chilika. pp. 96-103. In *Chilika, The Pride of Our Wetland Heritage* ed. S.N. Patro. Bhubaneswar: Orissa Environmental Society.
19. Ram, N. Rahul, Rama Rao, N.K.V., and Ghosh, A. (1994), *Ramsar Sites of India: Chilika Lake*. New Delhi: WWF India.
20. Chilika Development Authority (2009), *Annual Report*.
21. CDA. (2010), *About Chilika: Climate*. Available at http://www.chilika.com/climate.htm (accessed on October 10, 2010).
22. Sanyal, Sanjeev (2016), *The Ocean of Churn: How the Indian Ocean Shaped Human History*. New Delhi: Penguin.

2

Drawing in the Nets: Half Full, Half Empty

The familiar, stratified, choppy, unpredictable, bountiful or parsimonious waters of Chilika are perfect metaphors for the lives and livelihoods of the fishers. Like other human societies, the fishers of Chilika, are typified by prevalence and inescapability of social and economic disparity. Their endemic inequality is reinforced by the inequitable allocation of limited resources, which begin with and go beyond the interplay factors like status, position, power, ability, reverence and value[1]. Consequently, the certainty from grounding of social hierarchies is rooted in an understanding of these processes. Socio-economic conditions of traditional fishers of Chilika are also affected by the natural environment within its various ecosystems, which supplies goods in the form of fish, salt, and services like transportation and tourism. This implies that the stability of environment and the community welfare in these areas are highly interdependent. Socio-economic structures in Chilika may have a positive or negative impact on the various human activities conducted in the area resulting in changes in its ecosystem.

There are strong linkages between the causal factors deciding environmental vulnerability and the socio-economic crisis within the fisher population. The resultant debate is complex and encompasses both multidisciplinary scientific research and manifold theoretical arguments. The issue is not only about the different strata of society but the numerous dogmas and characteristics that exist within the society leading to different social groups and classes. Therefore, in the course of this chapter, as I evaluate the socio-economic dimensions of the fishers of Chilika; relevant conceptual perspectives, fieldwork, analysis and inferences will be integral parts of the narrative.

Conceptualizing the Socio-Economic Status

Social stratification often creates economic strata, by helping a group within a population to generate resources through the unique nature of jobs that they are allowed or can perform[2]. It is however important to understand the manifold attributes that affects the stratifications. Heterogeneous characteristics like family, religion, caste, age, education, occupation, language, health and hygiene play their individual, shared and multifarious roles in deciding the socio-economic status of an individual or family, making it either a resource rich or vulnerable entity[3]. Each of these aspects have been addressed in detail in this chapter and derived from in-depth fieldwork carried out *in situ* in Chilika and under the head "methodology mechanism".

In fact, Parson (1975) was the first to detail and reinforce the concept of socio-economic status from a theoretical perspective, even though Weber (1958) had long before conceptualized disparity based on caste, status and party. For Weber, each of these characteristics inherently carried associated power and influence; his ideas of class were more about financial capital and political influence, whereas status was more about reputation and respect; which was certainly based on a person's individual way of life, consumption patterns, and manners. Parson at a later stage meaningfully elucidated that status is a pillar in the social hierarchy, and is a part of social diversity, based on occupation, family position, etc. in the society; and thus carried the concept of Weber's notion of honour and prestige in the context of hierarchical reference.

In this conceptual frame, it is important to understand the different dimensions and threads of the socio-economic mosaic of the fishing community in Chilika, who are highly vulnerable to environmental degradations and natural calamities. Although the traditional fishers follow their instinct in building their socio-economic blocks, the entry of advanced scientific technology, has to a large extent impacted this system. Several studies have drawn significant attention to the socio-economic conditions of small-scale fishers. These findings reveal that livelihoods of small-scale fishers are plunged in poverty (FAO 1995; Rubinoff 1999; Sekhar 2004, Bene and Friend 2011). There is a growing recognition among experts that a fishery cannot be managed effectively without the cooperation and participation of fishers to make laws and regulations work (Pomeroy and Berkes, 1997). This participation does not have to remain limited to a community but must be integrated into larger democratic structures. The community should have a voice in policy level discussions and development debates, that influence their lives.

This does not mean that non-human nature in the present reading does not have an agency of its own, to affect and influence the way social customs are articulated. Most of the ecological dilemmas of our times are deeply rooted in social problems, an aspect persuasively asserted by Bookchin Murray (1964 and 2007) in his larger exploration of the concept of 'Social Ecology.' In our context the environmental issues that arise because of capitalistic modernity, corrosion of traditional beliefs, and centralized state structures, have found lucid expression in the work of Ramchandra Guha (1994). He delineates five important domains for social ecological interactions, namely, 'ecological infrastructure,' 'economy,' 'social structure,' 'polity' and 'culture.' The assumption that the differences are primarily based on the related skills vis-a-vis income would perhaps be wrong, as several other issues including gender, health and location, play a vital role in negotiating the fishers' current livelihood. The study of socio-economic situation of the fisher people, as presented under this section, is an initial attempt to locate Chilika within a larger framework of 'social ecological' interactions as mentioned in Bookchin (1964) and Guha (1994). This alone will pave the way for a more serious engagement with the issues at hand. Against these pragmatic principles and theories, FAO points towards even more practice-oriented notional ideas about fishers. It says, "the complexity of fisheries systems, characterized as they are by constant interactions between biology, ecology, economics and sociology, means that the economic performance of the sector depends both on the properties of the fish stocks, and on the market conditions and the efficiency of the harvesting process" (FAO, 2017). Globally there is limited data available on socio-economic issues of the fishery sector, partly due to lack of technical capacity for their collection, but mostly due to the complex and ever changing eco-system under which they function. However, as we discuss the traditional fisher community further, it would be important to highlight its evolving nature both in India as well as in the case of Chilika.

Who Comprises the Small-Scale Fishery Sector in India?

The story of the traditional fishery sector is one of gradual evolution and innovation of individuals, confronted with the option of making a living from fish. Small-scale fisheries are primarily caste-based and over time have evolved their own modalities in the form of fishing gear, crafts as well as fishing techniques for their survival. Over the years the indigenous knowledge of fishery management has helped the traditional fishers in developing sustainable fishery management, and simultaneously matching increased catch requirements in a highly competitive fishery business.

Figure 2.1: A fisherman inside Chilika Lake.

Kurien (1978) through a macro analysis of the fisheries development based on secondary data, analysed the trends in production and distribution of fish output during the post-mechanization period 1963-1976. He also claimed that the labour intensity of traditional fishers' technology, and the space limitations on their crafts, have restricted their operations to shallow waters nearer the coast, which abounds in large quantities of only the smaller species of fish. Thus, irrespective of the fact that fishing in an open resource area is a continuous pursuit of harvesting, inherent constraints, and opportunities that have arisen from equipment size and the perishable nature of the harvest. On the positive side, initial disposal of the catch and preservation are often the domain of the wives of fishermen. The initial preservation techniques like salting and drying increased the self-life of fish and further expanded the potential of fishing business. Subsequently men from other communities who did not consider the fish trade a taboo were also drawn in by the profitability and easy earnings. These post-harvest operatives left the fishermen free for further harvests and created more employment opportunities and strata within the community. This was till the new ice age dawned with the mass production of ice and its use in catch preservation that completely changed the traditional reach of the fishing business.

This technological revolution brought in its wake new cast and crew into the fishery sector to operate in packaging, storage, transportation and similar tertiary sector related activities. The technology enabled the fishery sector to vastly enlarge its scope and reach, which in turn required additional facilitators, who came in as middlemen between fishers and big businesses, providing initial finance support to traditional fishers. They initially took on the role of small traders, actively facilitating the sale of fish in the open markets. In due course, they started resolutely managing the fishery business at the source, including hiring their own crew with nets and fishing vessels. As Kurien (1978) has previously claimed, this class of middlemen-agents gradually began to exercise dominant power. This concentration of power also meant that practically all further expansion in productive activities depended considerably on them and the 'sharks' share' of benefits also accrued to them. They also ensured the barter-based system turned into a monetized system. The demand of fish slowly grew as a part of daily and changing dietary preferences, especially in the coastal regions, where fish became one of the crucial sources of protein intake. The traditional sector of the fish economy was made up of mass, rural-based sector of many producers, distributors and consumers. They ensured continuous supply of fishery resources even in the remotest location at affordable prices to meet the need of varied consumers. These producers were confined to pre-capitalistic production relations, used labour intensive technologies that were of low productivity, yet yielded proportionately large surpluses. The large number of small distributors undertook trade for subsistence and catered effectively to the widely dispersed and growing segment of low-income consumers. Therefore, though the main participants were from the traditional sector, but the actual control was with the finance capitalists, who gradually gained a crucial and vicious foothold into the production and marketing activities (Kurien, 1978). In line with this, this chapter covers the intrinsic socio-economic context of the fishing community under distinct yet interconnected headings.

Methodological Mechanism

The socio-economic impact assessment highlighted possible relationships between assessed variables and the environmental aspects in the study site. During the fieldwork efforts were made to recognize and understand the fishermens' perceptions, their relationships, realities, and the structures within which they exist; as the ultimate objective was not only to obtain the required data and information but also to involve them in the assessment process. Specific features of each fishing village were considered while listing all possible impact areas. To support the quantitative findings, I have further narrated the quotes of the respondents to support the arguments.

Keeping in mind the nature, scope and objects of inquiry, availability of funds, time factor and precision required for the sampling procedure, the method of individual interview with fishers and their community was adopted for eliciting information from the subjects. To locate sample areas and to identify the subjects, multi-stage-multi-purpose sampling method was adopted. Multi-stage sampling is a design in which two or more levels of units are embedded one on the other, for example, geographic areas (primary units), organizations (secondary units), and personnel (tertiary units). At each stage, samples of the corresponding units are selected, and this sampling method is sometimes used when no general sample frame exists. In this case, the first step was to select, at random, a sample of areas, collective units, or blocks (primary units) from a list wherein they were all registered. Then, for each selected primary unit, a comprehensive enumeration of all units of lower rank was done, thus obtaining a local sample frame, from which a sample of secondary units was selected. For example, for each village of the secondary sample, a list of all fishers' housing units was established, allowing for a selection of a sample of households. Different probabilities can be used at each stage, as well as within one stage, for the different units to be selected. Probabilities at the successive stages multiply, so that the resulting probability for selecting one final unit is the product of the probabilities used at each step.

As Chilika is surrounded by three districts of Puri, Ganjam and Khurda and the fishermen from these districts depend on Chilika, the study focused only on those blocks where the fishers reside. Out of these blocks, using a purposive sampling technique, three blocks from Khurda (Chilika, Tangi and Banapur), two blocks from Ganjam (Ganjam and Khalikote) and one block from Puri (Krushnaprasad) were selected. The selection of sample areas and identification of subjects were executed based on the following norms, like areas in which fisher folk resided exclusively or fishing activity is carried out in the periphery of the lake, proximity or distance from the immediate headquarters (Block) and villages having fishers of various categories residing within them. Utmost care was taken to envisage a balance of these norms. Each block has under its jurisdiction many villages, out of which 30 villages were selected randomly keeping the above norms into consideration. Further from each village, 15 households were selected for the study, totalling 450 respondents. The list of villages selected is as follows:

Table 2.1: Village, Block and District Profile of the Study Area

District	Block	Name of Village
Ganjam	Ganjam	Gajapati Nagar, Rambha, Subalia, Niladripur
	Khalikote	Langleswar, Pathara, Keshpur
Khurda	Chilika	Nairi, Balugaon, Baradihi, Bhabapur, Gangadharpursasan, Kalupada, Sananairi, Sorana, Baulabandha, Balia, Jarikuda
	Tangi	Tentuliapara, Chandrapur, Manalajodi, Bhusandapur
	Banpur	Galua
Puri	Krushnaprasad	Alupathana, Brahmapur, Khirisahi, Mahesa, Samantarapur, Arakhakuda, Satapada

The fieldwork was undertaken between 2008 and 2009. As I narrate stories from our fieldwork in the following chapters and put across fishers' opinions, quantitative data were further used to substantiate my arguments towards traditional fishers' vulnerability in the lagoon area. I must highlight that all our interviews took place in the homes of the respondents, though we also collected some data from the lake shore, where fishers usually gather before fishing and after fishing and a few local offices during office hours. In a few cases we followed up with some of our households to understand the problems they have been facing. During interactions with respondents, they also recalled different realms of sociality and different ways adopted to cope with the changing times. In several cases we were informed that their vulnerability is linked to their own societal failure, where the openness of resources and limited restrictions lead to deprivations and further marginalization. Also, at other times, the respondents stated that their narratives/livelihoods are much better than their fisher counterparts in other parts of the state, where the climate change and periodic restrictions in open areas were taking its own toll. Over time, I observed that winding together these dispersed explanations provided several conclusions to unfinished stories, evolving strains and linguistic outlines such as use of dysphemism or dire agony to express a range of issues through which mostly "*otherings*" is highlighted. Interestingly though some attributed such disorders or marginalization, as fate, others labelled them as normal transitions, with a hope for better days.

I must admit that most of the fishers we interviewed and interacted with were extremely gentle and many of them were eager to share their personal feelings and experience. However, a few were suspicious and remained silent during the interview as they feared political alignment on our part. Some of the fishers initially refused to share their dedicated time and insisted we travel with them in the boat when they were fishing to learn of their views. The questions were asked in Oriya, which is their lingua franca and our mother tongue. Only in one village, where there was a dominance of Telugu fisher people, we faced some problems in terms of communication. However, with the support of a local youth, this problem was addressed. Also, during the interactions, I tried to understand the causes behind the conflicts among the fisher communities; however most of the respondents remained silent on this issue. While some of them had apprehensions that their views could be taken as personal, which may go against them; others did not want to remember those incidents. It was only during the interviews that I realized many of the fisher people had very distressing experiences during such conflicts.

During the fieldwork, we also interviewed a wide range of stakeholders including government officials, leaders of various students' organizations, community leaders, functionaries of civil society organizations and academic organizations. We also covered a few case stories and focused on group discussion with the fisher community members. In addition, we extracted a number of documents from government offices on various issues, especially on fishery statistics for our reference.

The Narrative in a Nutshell

The first chapter of this book sets the overall context and locates the fishery sector among multiple stakeholders, control mechanisms, global and regional contexts. In the second chapter, geography, ecology, potential versus disruption of livelihoods, global dynamics of the sector regulation and socio-economic implications frame the narratives cardinal to sustainability in fishery communities. The third chapter highlights the complex relation between several variables like caste, gender, age, patterns of resource use, health and hygiene in the context of fishery livelihood in Chilika. Using these vital social aspects as an anchor, we have been able to articulate how fishers perceive, act and navigate their well-being in the context of decreasing fishery resources. Consequently, we have evaluated poverty in the context of declining of resources, rather than the absence of entitlements, which would require inclusion of violation of rights, and necessitate further exploration. The fourth chapter highlights, how conflict around the lake has gained its momentum based on the perceived rights of commons. Fishing in the lake, follows an age-old system of sharing boundaries which is not easy to comprehend or visualize. The traditional access to rights in the lake has evolved over time and is followed by the communities to manage their fishery resources sustainably. However, over the last few decades, developments in the fisheries sector has led to the traditional fishery-related activities being usurped by a new group. The central argument of this chapter reflects, the emergent conflict created by lack of protection to the locals and absence of guidelines by the state, to claimants of user rights over the common. The fifth chapter plots livelihood patterns of fishers and enquires into some of the external and internal variables of land ownership, possession of livestock, market attribution, shrimp export, that impact fishery livelihoods. The chapter specifically highlights how the fishers have been subjugated by the combined external forces of middlemen and moneylenders, who remain inexpugnable, despite the presence of formal institutions. Putting it all together, we argue for a more regulated credit structure in the region. Chapter 6 deals with the issues related to the marginalized fishing communities who are victims of and vulnerable to environmental hazards and livelihood disruptions brought on by new equipment, mechanization and illegal fishing. After taking stock of the conflicts revolving around the mechanization of fishing vessels and the alienation of artisanal fishers from traditional fishing, we have argued that the issues of mechanization and ecological hazards have their own bearing, on not only livelihood disruptions but also the *in-situ* footprints of common property resources. The last chapter deliberates the combined adverse impact of climate change and environmental degradation on the bionomic based lives and livelihood of traditional fishers. We assert that casual and ad hoc redressal of disaster risks cannot yield sustainable measures to combat the extant vulnerability from repeated disasters. Consequently, there is a pressing need to view safety and livelihood with an integrated lens that draws socio-economic challenges into its field of vision.

NOTES

1. Davis (1942) highlighted that disparity in terms of access to the basic resources of the society has a dimension of prestige and honour which acquires various layers by virtue of who they are or their position. Consequently, among different strata, social inequality is both positively functional and inevitable based on the amounts of scarce and desired goods, and the extent of stature they maintain and the reverence they receive.
2. In order to ensure a functional and serviceable community structure, society distributes its members based on social positions and ensures that they perform the duties of these positions. As societies evolve, stratification persists even when the composition of individuals within different strata undergo continuous mobility on acquiring skills to perform newer roles, interspersing socio-economic differences as a result.

3. Parsons (1975) defined "family units" as one of the important stratifications of the social construction as it is presumed to be the unit of cohesion sharing similar well-being and had a single breadwinner and was central to his understanding. Similarly, religion plays an important role in socio-economic status as rituals are based on credence and morals that are represented by tangible sacrosanct objects, which in turn is connected to an eloquent way to the evidences of the individual's life.

REFERENCES

1. Davis, K. (1942), A Conceptual Analysis of Stratification. *American Sociological Review*, 7.
2. Parsons, T. (1975), The Sick Role and Role of the Physician Reconsidered. *Milbank Memorial Fund Quarterly*, 53, 257–78.
3. Weber, M. (1958), *The Religion of India: The Sociology of Hinduism and Buddhism*. Gerth HM, Gelncoe Don, translators. Ill: Free Press.
4. FAO (1995), Code of Conduct for Responsible Fisheries, Rome. www.fao.org/DOCREP/005/v9878e/v9878e00.htm.
5. Rubinoff, Janet Ahner (1999), Fishing for Status: Impact of Development on Goa's Fisherwomen. *Women's Studies International Forum*, Vol. 22, No. 6.
6. Sekhar, Nagothu Udaya (2004), Fisheries in Chilika Lake: How Community Access and Control Impacts their Management. *Journal of Environmental Management* 73.
7. Béné, C., and Friend, R.M. (2011), Poverty in Small-Scale Fisheries: Old Issue, New Analysis. *Progress in Development Studies*, 11(2), 119–144.
8. Pomeroy, R.S. and Berkes, F. (1997), Two to Tango: The Role of Government in Fisheries Co-Management. *Marine Policy*. Vol. 21, No. 5.
9. Bookchin, M. (1964), Ecology and Revolutionary Thought, Comment (republished in the British monthly *Anarchy* in 1965).
10. Bookchin, M. (2007), *Social Ecology and Communalism*. Oakland, CA: AK Press.

11. Guha, R. (ed.) (1994), *Social Ecology*. Delhi: Oxford University Press.
12. Kurien, J. (1978), Entry of Big Business into Fishing: Its Impact on Fish Economy. *Economic and Political Weekly*, September 9, 1978.
13. FAO (2017), *Handbook for Fisheries Socio-Economic Sample Survey; Principles and Practice*, Rome.

3

Analysis and Inferences: Relocating Kalijai

Woven into the basket of narratives that we recorded, was the poem on Chilika called "Kalijai" by Pandit Godabarisa Mishra. It was a childhood memory, that kept coming back to me and was often quoted by the fishers. The story is about a child bride named Jaai, who when she came of age, as was the custom, had to leave her home, and travel to her in-laws' house in another island named Parikuda. On the eventful day, as Jaai was being taken by her father in a boat across Chilika, the poet describes Jaai's extreme agony at the loss of her childhood and the inescapable bondage of womanhood. As they navigated the waters of Chilika, a tremendous storm and whirlpool arose, overturned the boat, and carried away Jaai. No amount of searching by the boatman and her father could even bring back her body. But generations of fishers continued to believe that Jaai's voice could still be heard in the island adjoining the whirlpool. Overtime, she came to be venerated as Goddess Kalijai; a homage to their belief that Jaai had been reincarnated as a form of the Hindu Goddess Kali. The Kalijai temple was built by King Jagannath Mansingh of present-day Banapur in the year 1717. Even today on the way to the temple, the boatman takes a circuitous route to avoid the area where she disappeared. Given the nature and parameters of the investigations into the socio-economic aspects of the fishers and their livelihood in Chilika, Jaai's story is oddly apocryphal. It almost seems to warn us of unknown depths, ungovernable weather, fragile lives of the fishers bound by tradition and the waters of Chilika. Even from our interaction with fisher people, we learnt that they believe the goddess Kalijai serves as an agent of the water and its resources, which in turn signalled the presence of strong socio-cultural elements that govern their lives, livelihoods and traditions.

Therefore, into our records, the voices of the fishermen have been given due credence and quoted in the context of the societal, cultural, individual and economic aspects being analysed. The data collected as previously detailed in Methodological Mechanism, has been analysed under 10 individual heads; Religion and Culture, Age and Intergenerational Dynamics, Caste and Tradition: Patterns of Resource-Use, Education, Language: Official and Traditional, Families in Transition, Marriage and Gender Related Responsibilities, Income Earning and the Sharing of Labour, Health to Housing and Hygiene.

1. Religion and Culture

The significant impact of diverse religions on Indian culture has been well established by T.N. Madan (1993; 2004), one of the founding fathers of Indian anthropology. Religion plays a key role in the lives of people in Odisha and in the lives of the fisher people living in the Chilika region. Understanding the religious basis of the opinions held by fishers, helped in elucidating the role of religion in fisheries management, and the socio-economic management processes. In a broader context, understanding how religion influences values and behaviour of the people, becomes an important facet of environmental conservation. Out of the total respondents, 98.4 per cent follow Hinduism and 1.6 per cent follow Islam. The Indian Census data of 2011 shows the state of Odisha is predominantly Hindu and the proportion of Hindus to total population is 94 per cent. The villages adjacent to Chilika have a dense Hindu population and this majoritarian uniformity shows an observable degree of parity in culture and traditions. This is not to say that religion acquires a profile of homogeneity in the region, yet a similarity in style of managing families and common customs can be observed. Fishers of Chilika hold strong Hindu beliefs which significantly impact their business, and throughout the course of our interaction with fishers, we were told that Hindu ethics plays a significant role in shaping the fishers' attitudes towards fishery. In public meetings, socio-economic discussions surrounding the fishers often turn into religiously charged arguments. Younger fishermen who are not equally oriented towards indigenous belief patterns frequently withdraw from or dismiss religious arguments, thus shutting off communication with the elders. The emerging gap between scientific and indigenous knowledge systems has received its fair share of attention from commentators on the subject (Gadgil and Vartak, 1975; Rangarajan and Saberwal, 2003; Sahabuddin 2010).

Figure 3.1: Black and red bangles hanging from tree branches in Kalijai, Chilika. The custom of hanging bangles is practised by tourists and villagers desiring for their prayers to be fulfilled.

Yet, in the case of Chilika we notice something more perverse: indigenous wisdom is not only being alienated from the so-called 'modern' and 'progressive' forms of environmental thought, but charged with its own share of internal differences on religious grounds. In most of the cases the fishers' disagreements on management of fishery resources is based on powerful and sincerely held religious beliefs. For example, there is a strong belief among the fisher people that Goddess Kalijai provides fish for the sole purpose of being harvested. Clearly the goddess Kalijai is imagined as a picture of benevolence and not a destructive and ferocious manifestation like some of the early Rig Vedic gods; further the fisher people believe that the goddess serves as an agent of the water and its resources. During preliminary discussions with local NGO officials, we were informed that the fisher people are adequately trained on natural disasters and strictly follow the safety guidelines while fishing. However, the local priest informed us that most fishermen depend on their instincts and strong religious beliefs while going out for fishing. A leaf can be taken from Annu Jalai's celebrated work, *Forests of Tigers*, where she demonstrates the relevance the goddess Bonbibi holds for the local population living in the harsh climatic conditions of the Sundarbans. Jalai (2009) ponders the long history of the role of traditional beliefs, especially those furthered by the *pirs* and *Sufi* saints, in creating a sense of safety and security for humans in hostile environmental conditions. The values of such traditional beliefs were recognized by British officers such as Francis Buchanan in the early part of the nineteenth century. Unfortunately, Jalai shows how Bonbibi has become an active site for the domination of the more elite sections of society, who seek to redefine the God in a more urbane way. We do not observe any displacement of such traditional customs in the case of Chilika and Kaljai. The goddess Kalijai has retained her largely proletarian following far more successfully, despite the earlier mentioned differences in opinions. From the religious point of view the fisher people strongly believe that Maa Kalijai (Mother Kalijai or the Mother Goddess), the local deity, decides their birth and death. Besides, the fishermen as well as the non-fishermen from the villages brave fishing even when high velocity winds rage, because they are sure that their destiny, survival and suffering is guarded by Maa Kalijai. The island of Kalijai is consequently renowned as a centre of spiritual devotion due to the temple of Goddess Kalijai; and a big fair is held on the occasion of Makar Sankranti that falls during the second week of January every year. Empirical studies from McCleary and Barro (2006) reveal a negative correlation between religion and economic performance. Conversely, we found a strong association between the religious beliefs and socio-economic status of the fishing community in Chilika. The central point of our argument is that while religion affects important economic outcomes for the fisher community in Chilika, conceptually and empirically measuring its prevalence and significance is a complex process, as religious beliefs among the fishing community are multidimensional and have a strong linkage with the age of fisher people.

2. Age and Intergenerational Dynamics

Biologically, age indicates physical and mental well-being. However, sociologically age is an important factor that not only determines one's social status, but also economic status. Both theoretical models and empirical results suggest a strong age and wealth relationship (Davies and Shorrocks, 2000). There is clear evidence that elderly people experience added disparity and inequality in the context of economic, ecological, or social domains. On the other hand, younger populations are less vulnerable to both economic and social aspects. For the fishery sector, age plays a critical role, as only the younger and able persons match the rigours of fishing in open waters with their robust physical abilities.

As far as fishing as an occupation in the communities around Chilika is concerned, the study reveals that a significantly higher proportion of people in the 31 to 45 group age are involved in fishing, compared to the 15-30 year age group. We also observed that fishers quit or fish less after fifty years of age. This may be due to the hazardous nature of the job and stressful life that forces the male earning members of the fishermen's families to give up fishing at a relatively early age. It is important to remember that the risks inherent to the fisheries-based livelihoods are not simply because of the 'natural' disasters, but equally the social contexts in which they occur (Wisner et al., 2003), is a theme that I explore separately under a separate chapter of this book. It is also true that there is no alternative earning opportunities, livelihood or occupation for fishermen over 50 years of age who have quit fishing; and are thus forced to stay at home. This leads to a situation of financial destitution in most of the cases and in turn increases the levels of economic risk, leading to a vicious cycle, where both the natural perils of a livelihood and adverse social contexts compound one another (ibid). Irrespective of this situation, the older fishermen retain a strong relationship and attachment with the lake. Mr. Padhiari from Balugaon reflected upon this intimate attachment with the lake.

> I have no skills or experience of any work other than fishing. My parents and their friends have taught me how to get the best out of water; and till the end of my life I can only use my fishing skills for survival. Water is our life as it gives us earnings. In the last few years our kids are making more money from fishing, irrespective of the fact that there are now so many people who depend on fishing and you can see the big boats mainly make good business. We will never move from here, or ever leave fishing, till the time the lake stops giving us anything.

This statement not only conveys a sense of confidence, but also a feeling of security and assurance. This innate belief in their potential ability to resurge even under difficult circumstances further reflects the strong relationship between people's resilience and dependence on the lake water. Mr. Padhiari's statement reflects that his understanding, relation and adjustment with the waters of the lake are long term and unrestricted[1]. To comprehend the relationship between livelihoods and issues of poverty in the context of fishers, one must understand the traditional livelihood arrangements and how they have been managed over the course of time. At a broader level, policies, institutions, and processes affect the livelihood of fishermen on a much larger scale, and curate the vulnerability of the environment in which fishers live and work. Fishing being a seasonal occupation, a fisher's vulnerability is also dictated by the level of availability of other economic activities in the neighbouring areas.

Allison (2001) suggested that entry and exit into a fishery is more related to income levels of fishers relative to other sectors. However, in an East Africa lake fishery, Ikaria and Odin (2000) concluded that fishers remain in their occupation, not because it is less profitable than other sectors, but due to lack of alternatives. They found that the fishers' resistance to exit the fisheries was related to the opportunity cost of exiting, fishing experience and vessel ownership. Pollnac et al. (2001) also demonstrated that non-economic factors influence exit decisions. Their study cited the pleasurable aspects of the occupation and tradition as some of the non-economic reasons for fishers to resist occupational change. However, those willing to change occupations had specific characteristics in each country. Our study shows that 85 per cent of the respondents continued with the fishing as an occupation for the last four or more generations, 6.9 per cent for three generations, 6 per cent for two generations and 2.2 per cent for their current generation only. This proves that fishing is the most prevalent and vital occupation for the people living in the Chilika region. In the recent times with the advent of modernization and mechanization of fishery-related activities, several traditional activities that used to be undertaken by the fisher people have now been taken over by the new generation business groups, while others have become redundant. However, this is not true in case of the fisher people of Chilika. They continue to follow the traditional approach of fishing as an occupation. Jiban Dash from Balimara village narrated his understanding of vulnerability as a fisherman.

> We are not planning to move from here at any point of time, though I am fully aware of the risks. In my life, I have seen acres of land being swallowed and several acres of land coming out of water due to erosion as well as sand deposition. I have seen several clashes between our own people; and because of this lake, my house got severely damaged during the cyclone, though fortunately we did not lose any of our family members. This is our land and till the time Maa (Kalijai) wants, we will survive and live on fishing.

For the fishermen, earnings from fishing business leave very little surplus beyond their subsistence needs. As a result those who retire from fishing on account of old age find themselves at a loose end. Even households that generate some surplus use it up quickly during lean periods or spend on repairs of boats and houses or for festivals. Those who depend only on fishing, get into debt traps as soon as the savings are spent. This needs to be looked into in line with the depleting fishing resources and environmental degradation. It is also true that fishers were quick to adopt new technologies when required and simultaneously, new technologies have considerably eased operations. Therefore, as an alternative, action needs to be initiated to establish new business options for the traditional fisher people belonging to older generations.

3. Caste and Tradition: Patterns of Resource-Use

Ninian (2008) highlights that about 180 million to 220 million people are considered to be in the lowest caste in India and these lower castes are basically alienated (Thekaekara, 2005). Caste assigns the ascribed social status to its members and opens occupational prospects for the members. The caste system in India is deliberated not simply by the disputing arguments of hierarchy, but by concerns of disparity and scarceness of resources among individuals; as it could be a significant group identifier in matters relating to economic inequality and subjective discrimination. Shekhar (2004) however brings an important argument that India is one of the best cases to demonstrate a situation where the predominant presence of a caste system is deeply entrenched in society and thus defines the status of a fisher.

While trying to understand the castes among the fishermen; it was interesting to learn that the fishermen of Chilika acknowledged seven castes in the area, against the prevalent differentiation based on Scheduled Tribes or Scheduled Castes (reflected in the following Table 3.1). Based on which we tried to further understand their caste background, to determine the achieved social position in accordance to their occupation.

We observed that the Keutas are the most dominant caste among the fisher people of Chilika, followed by the Kandaras and Tiaras. A clear caste-based demarcation in relation to fishing emerged out of our interaction with the fishers. The Keutas fish in deep waters of Chilika with the help of cast nets, and they dominate the fishing community. The Kandaras fish using different kinds of traps, while the Tiaras fish in shallow areas in Chilika, using Dhaudi, Baja, Tata[2] for catching fish and shrimp juveniles. These people do not use nets. The Nolias mainly catch marine fish near

Table 3.1: Caste of Fishermen in Chilika

Sub-Caste/ Block	Ganjam	Khallikote	Chilika	Tangi	Banpur	Krushnaprasad	Total
Keuta	17	31	108	37	0	64	257
Kandara	38	14	0	0	0	7	60
Tiara	0	0	23	0	15	6	44
Nolia	5	0	0	0	0	21	26
Niary	0	0	5	1	0	0	6
Karita	0	0	0	2	0	0	2
Khatia	0	0	6	4	0	4	14
Others	0	0	24	0	0	2	42
Total	60	45	165	60	15	105	450

the lake's mouth with big drag nets and cast nets. This segregation of resources among different caste groups points to the relevance of Gadgil's thesis on 'resource fragmentation. By pointing out how the Phasepardhis, Vaidus and Nandiwallas, in western Maharashtra, had all specialized in hunting different animals. Gadgil (1999) remarked that 'competition for limited resources through diversification of resource use and intra and inter-caste territoriality' is a remarkable regulatory mechanism of the caste-based society. Thus, even though caste is largely divisive nature, its dismantling over time does not always augur well with social arrangements over water and its accompanying resources. Therefore, caste-based occupational as well as livelihood arrangement in Chilika should be examined far more in depth.

Benu Behera from Ganjam narrated:

> Caste system in the fishing community is difficult to define. Within Keuta there are two groups, those who make beaten rice and the others who fish. With the growing mechanization the groups of Keutas who used to make beaten rice are now getting into fishing, even though earlier they considered Keuta fishers to be inferior. Similarly, there are several upper caste people, like the Kayasthas who have now got into the fishing business, due to the demanding occupational rigours and finally there are many low-class people like the Telugus, who have also moved into the fishing business.

This throws up an interesting lead, which indicates that a small but clear caste-based upward, downward, and horizontal mobility has taken place in Chilika, primarily due to demanding occupational rigours and the open nature of the lake.

During the Focus Group Discussion, we were told that today, none of the castes follow any specific method for fishing and fishers keep changing their methods for maximizing their catch. Thus it is evident that caste-based positive regulation is gradually taking a back seat; therefore its key role in deciding the fishing methods, which had a major bearing on the fish catch and a direct bearing on the income is fast losing ground. While absence of caste-based arrangements are generally hailed as positive; they simultaneously, demand a careful re-assessment of the fish stock at hand and the long-standing communitarian arrangements that have regulated the harvests. Consequently, these changes require urgent measures to protect sensitive areas such as the outer channel of the Chilika Lake, Magarmukh channel, and Palur canal. They should be prescribed as "no fishing zones" and use of any type of nets and Khanda in these areas needs to be strictly prohibited through institutional measures that are binding.

Against the common notion that the lower caste people are more likely to be engrossed in low-paying and less-prestigious occupations, we found that within the fishing population, castes play no such deciding role. The overall social and economic suppleness is still a distant dream for the fisher people; and there is a strong link which leads to the belief of the old Marxian judgment of a caste-class overlap. As I argue for the growing incidence of the breakdown of the traditional subsistence economy among the fishing community, and that the influence of caste is not snowballing among the community; I clearly perceive that with growing mechanization and increased fragility, the caste-based manifestations are changing. However, it will be definitely important to look at the economic system of Chilika from the perspective of a class-based economy, as the fishers' notion of the caste system varies from that of the prevalent Indian caste system.

Caste has a strong relation on the traditional livelihood pattern in India. Thus, the rural-urban migration of villagers induced by a search for alternative occupations due to the agrarian crisis has impacted these traditional livelihood systems. During the study we also delved into the question of which occupations had been traditionally taken up by the respondents apart from fishing.

Some of the older respondents informed us that the fisheries of Chilika were part of the zamindari estates, lying within the kingdoms of the Rajas of Parikud and Khallikote. The zamindars used to lease out the fisheries exclusively to the local fisher folk. The British also started a cooperative store in Balugaon in 1926 to provide fishing equipment to locals. After the abolition of zamindari in 1953, traditional fishing areas continued to be leased out to cooperatives of local fishermen. Thus, Chilika speaks of a rich and long-standing mix of traditional resource-use and emergent institutional structures. Most of the respondents in the study, i.e. 94.6 per cent were involved mostly in the fishing sector, about 2.4 per cent were engaged in cultivation, 2.2 per cent in the service sector and a negligible 0.2 per cent in trade and business and 2 per cent in other activities. While fishing is the main occupation of the respondents, it is anticipated that with time as fishery resources decline, there will be a pressure on the traditional fisher people to switch over to new occupations, leaving behind the traditional occupations. There is an urgent need to evolve strong socio-institutional mechanisms to deal with such changes. Such mechanisms would have to be alive to the dynamics that have emerged throughout our discussions.

4. Education

Education in India is constrained by socio-economic conditions of the people, their attitudes, values and culture. The Constitution of India 86th Amendment Act, 2002, inserted Article 21A in the constitution; it provided for free and compulsory education for all children in the age group of 6-14 years, in an appropriate classroom near their neighbourhood as a fundamental right, in such a manner, as the state may, by law, determine. With the enforcement of the RTE Act in 2010, education has become a fundamental right and an entitlement of every child.

Fishing communities are often remotely placed, suffer from lack of infrastructure, and are also marginalized from government development initiatives (Allison, 2003). Children growing up in such places, in common with other rural areas of developing countries are often deprived of education due to remote access to formal education, compared to the urban children (Meinert, 2003; Ansell, 2008; Lewin, 2007). The demands of the fishing industry, the strong occupational identities and incentives to remain in fishing are thought to negatively affect the educational outcomes of formal schooling in these contexts (Maddox, 2006). For a fisherman, it is important to be literate to manage his day-to-day work including measuring his catch, selling at the correct price, dealing with moneylenders, etc. Hence, a deeper, more nuanced appreciation of educational status is important for understanding not only the intergenerational transfer of poverty, but also the transference of poverty over the course of a lifetime. For the convenience of research, we identified the sampled fisher people into three groups: illiterate, barely literate and highly literate. In terms of reading and writing only the local language Oriya was considered. For the literate, we considered those who could read and write fluently, for barely literate we included those who demonstrated reading and writing with some difficulty and illiterate meant those who could not read or write anything at all.

The educational profile of our respondents showed that 3.6 per cent were literate, 73.8 per cent barely literate and 22.7 per cent of the respondents were illiterate. This is an issue and challenge for the development of fishers in the region. It proves that currently, attainment of formal educational in fishing communities is very low.

For the fisherman, it is important to send their children to school, as this may help to ensure their long-term livelihood and ensure alternate livelihood in times of crisis. After attaining the minimum qualifications, the new generation can also opt for a new source of livelihood, distinct from their family-based business. The study shows 89 per cent respondents wanting to educate their children, reflecting their awareness of the importance of education; and 10.4 per cent respondents replying in the negative. The reasons behind the situation are discussed in the previous section. Exposing children of fishers to life skills and quality education can help protect them against vulnerability to poverty, abuse or exploitation and enable them to prosper later in life. With the current situation in Chilika, which requires more community participation and effective governance, the new generation of educated fishermen will have better qualities of critical thinking skills, healthy living, resilience, and self-confidence.

Use of incentives, is a prime area to be considered, especially with regard to flexible school attendance schedules, which could offer a more effective route to train the children of fishers in several other areas. Such incentives could accommodate possible seasonal fluctuations, when children are required to help their parents in household work and even fishing, as suggested by some fishers during our interactions. Since fisheries related work is physically demanding and cannot therefore be learnt before a certain physical maturity is attained, there is a window of opportunity for using the consecutive gap between the ideal times for cognitive development. This will ensure that the rest of the parents, who do not feel the importance of education for their children, will send them to schools. Consequently, it implies implementation of child-friendly schools with policies aimed to improve the quality of schools with better infrastructure and good learning materials.

Attainment of education lays the foundation for greater overall economic productivity, and the full use of new technologies for development. The key parameters on which data was collected, centred around the reasons why fishermen want their children to be educated; they ranged from an improvement of social status, or doing better in the traditional family fishing business to pushing the children into other occupations in trade, the government, private or corporate sectors. A sizeable proportion of respondents (57.8 per cent) want to educate their children to push them into other occupations like government service, services in the private or corporate sector and trade and business other than fishing; and a mere 9.6 per cent want their children to carry out their traditional occupation better. About 22.2 per cent feel that education will help them to improve their status in society, where as 10 per cent had no answer. The responses to the queries reflect the increasing change in the attitudes of the people towards the acceptance of education; and as for fishing, they would rather retain an intergenerational occupational continuity for their children.

Although the literacy rate in villages continues to be low, the study reflects that many fisher people have begun to take an active interest in education and are sending their children to school. As mentioned above, the main reason for getting children educated is for them to get jobs elsewhere. This reflects that the fishers are aware of the current market situation and at the same time know that the decrease in fish catch will continue. Children's education will open opportunities of employment for the next generation, as most of the educated youth may opt for a job rather than fishing. Those who do not opt for jobs elsewhere would also look for alternative livelihood options within their village. Education and awareness will enhance their 'capabilities' or alternatively their 'functioning' (Sen, 1992, 1999), be it in their interactions with local traders and merchants or bureaucrats and policy makers. It will also improve their business ability by bringing in innovation and advanced technology in the fishing sector for better volume of fish catch; besides strengthening their bargaining power to secure higher market value for their products.

5. Language: Official and Traditional

Language is a part of tradition and the mother tongue of a child is part of her/his personal, social and cultural identity and has a direct relation with her/his education. The importance of the mother tongue has been emphasized at several forums on language and education as they are key factors for education and the success of other development efforts. It was in the 1971 Census, that the mother tongue was defined as "language spoken in childhood by the person's mother to the person". Nettle (2000) argues that the role of language in economic development is inconclusive. Oriya is the mother tongue of the people of Odisha and so also is supposedly considered to be the mother tongue of the people living in the Chilika region. But some migrant fisher people from West Bengal and Andhra Pradesh also speak Bengali and Telugu. During the study, it was observed that out of the total respondents, the mother tongue of 90.1 per cent is Oriya, that of 3.6 per cent is Bengali and that of 5.6 per cent is Telugu. As a medium, language plays a vital role in constructing and maintaining traditional ideas of nature. The local fisher folk have different terms to denote the movement of the tides, the quality of the water, and species that are found in Chilika Lake.

The Telugu fishermen fish near the lake's mouth and catch mainly marine species. None of them could tell us when they arrived here and for how many generations they have been here. But it is quite clear that none of them are waiting for an opportunity to return to their village or state. Many of them also maintain a surname of "Telugu". Rama Telugu told us his story of how he feels more secure here:

> Perhaps we are much better off here than we were in our homeland. Here we have never experienced hunger of which our grandparents used to narrate. We get fish throughout the year. Even if we do not get fish during the lean season, many of us get work in something or other. Some of our younger generation boys have gone back to Hyderabad (state capital of Andhra Pradesh) for work. But our daughters have been married here.. We are happily living here with no dearth of food and we are not going anywhere. More importantly in a different country (in Odisha) we feel secure with our Odia bhais (brothers).

However, it is highly plausible that the local fisher folk have terminologies that are perhaps missing or absent in a Eurocentric taxonomy. Commonality of language makes transaction easier and reduces the chances of conflict and misunderstandings. Different languages sometimes create differences between people. It can also create differences between different stakeholders who interpret nature, environment and ecology in different terms. Thus, language encompasses just about every facet of individual and social life, including economic processes. However, the impact of language on the fishery sector is extremely broad and the kind of link that we want to establish between the two is difficult to locate with accuracy.

6. Families in Transition

Families are microcosms of complex social arrangements that include caste, tradition, culture and linguistic backgrounds. In India, the family structure can be categorized into the joint and nuclear families. An extended, joint or undivided family normally consists of many generations living under the same roof, whereas a family consisting of parents and their children is called a nuclear family. Against the traditional and classical model of thinking that family does always act as a union and no conflicts of interest arise, contemporary studies are focusing more on how resources are distributed among the members and how family members constantly negotiate limited resources. To define and measure the standard of living, it is necessary to present a comparative description of the family structure of fishers of Chilika. The family structure of fishermen has definite consequences on the attitude and behaviour of the family members at the micro level. Depending on which factors are found to contribute more to the family income, the government can also design its interventions and policies so as to achieve the maximum effects from a given level of public expenditure. The study reveals that 68.2 per cent stay in nuclear families and 31.8 per cent still maintain traditional joint families. The Indian joint family system is changing and today there is constant progression towards a nuclear kind of family system. It is clear from the data that there is a predominance of nuclear family structure as against the joint structure.

Though large families remain significant in the fishermen's community, it is no more a dominant component of their lives. The life of fishermen being stressful and hectic, they look for a more comfortable life at home and the nuclear family might help them to avoid stress and discomfort. It also encourages initiatives, independence and self-reliance among the fishermen. Both nuclear and joint family systems speak of their own share of strengths and vulnerabilities, and the downside of emerging nuclear families is that they lack the social or cultural capital of earlier times. The emergence of nuclear families, especially in rural India, is not always accompanied by the presence of a more responsible and enabling state or government machinery. This frequently forces members of nuclear families to rely on moneylenders, contractors and market arrangements that are exploitative. In other words, social security provided by extended members of the family is eroded.

The study proves that an important function of cultural transmission performed by the family is affected because the nuclear families are scattered cultures. However, with the existence of such a large number of joint families, the study also proves that joint families are also favoured by the fishermen, because the higher number of family members provides more hands for labour which helps in more fishing and harvests. A high proportion of joint families could also possibly indicate the traditional nature of the activity and the lack of significant occupational mobility required for an alternative livelihood. As seen in the fisher families, while the fathers remain outside the home for most of the time, the mothers are engaged in household work. Prahlad Behera from Mahisha village was of the opinion that:

> With the growing pressure on economy, teenagers do not get a chance to share their emotions with their parents and become over-impulsive, which in turn makes them unable to control their frustration sometimes. The decrease of emotional bondage between parents and children to a large extent may result in a communication gap in future, unless it is handled in a more systematic way.

It was only in the 1971 Census that a household was defined as 'a group of people who commonly live together and would take their meals from a common kitchen unless the exigencies of work prevented any of them from doing so'. For the census, the important link in finding out whether it is a household or not, is a common kitchen. There may be one-member households, two-member households or multi-member households. Census 2011 reflects that 24 per cent households in India have a household size of 4 and almost 43 per cent households have an average household size of 4-5, while 25 per cent of households have 6-8 members in the family.

The size of their families plays a key role so far as their income is concerned. The bigger the size of the family, the more members are involved in fishing, leading to more income. The family size of fishermen is influenced by multiple factors such as economic, socio-cultural, environmental and education. Most of the time the family size of fishermen, affects their basic needs, as well as the income, growth, and savings.

In terms of family size, 24.0 per cent of the respondents had 3 members, 67.1 per cent had 4 to 6 members, and in case of 7.8 per cent it varied from 7 to 10. Slightly more than 1 per cent of households had a family size of 10. Thus, the number of family members of maximum respondents varies from 4-6 and that the very least (1.1 per cent) is more than ten. The small family size of the fishermen may be measured on the grounds of the ability to effectively cater to the needs of the family with a level of relative security. The family size of a fisherman may lead to low levels of income resulting in poor education, health, welfare and economic status, especially when the older fishermen in the family are not able to fish. To ensure a better social as well as economic standing, the fishermen prefer to choose a family size that would lessen the burden on the family as well as the individual members. Some of the respondents from the larger families indicated a relationship between the size and the income of the family. The larger families with younger age group members are found to have higher income, with one of the contributing factors also being supplementary incomes from other sources. Presence of more members within the family provides the opportunity to be engaged in other occupations or enterprises other than the main source of livelihood of fishing.

Our study reflects that a combination of high stress with social isolation for families can possibly result in many forms of dysfunctional family outcomes and therefore should be addressed with due care and consideration. If a family experiences multiple kinds of stress and 1) they are socially isolated and emotionally disconnected from one another, and 2) they are depressed and disempowered, then they will be at increased risk of illness, accidents, child abuse and neglect, and substance abuse, delinquency and school failure (Belle, 1980).

Family members engaged in fishing have a major impact on the overall family growth. This component was studied closely to understand the livelihood options for fishermen. The earning or capital in any form available with an individual or a family is neither fixed nor equal between individuals and families. The amount of capital available with individuals (households) is awarded or acquired by family members. Individuals (households) differ in their income-generating opportunities and access to the consumption smoothing channels. Different types of occupational engagement of individuals and varied number of earning family members, contributes to the economic position of a family or an individual.

During the research, we made an attempt to understand how many members from each fishermen family are engaged in fishing activities. Over dependence on a single source may make the families vulnerable, as with the decrease of fish catch, it may put undue pressure on the whole family. While in most of the families, only one or two people are engaged in fishing, in some cases six and above members of the family are also engaged in fishing activity. In 36 per cent cases, one member of the family is engaged in fishing, while in 35.5 per cent cases two members of the family are engaged in fishing. In 17 per cent cases, three members of the family are engaged in fishing and in 6 per cent cases, four members are engaged in fishing. Only in 4 per cent of the cases, five members and in 2 per cent cases, six members are engaged in fishing. It was also observed that in some families, the members are slowly getting into other businesses. Most fishermen have opted for other sources of income-generating work as unskilled labour, in neighbouring villages or urban areas through agents. Mr. Behera from Mahisha further narrated that

> The youth are migrating as labourers, and elderly people work in villages, since labour work is too hard for elderly people. People are also getting into non-fishers activity including agriculture business. With the different number of family members involved in fishing activities, each family or individual has a different economic status. Thus, for families or households which have low income resulting to low financial investments, there should be schemes to provide goods and services at low costs, insurance schemes, etc. This is more important for the older fishermen, those who are skilled in fishing but have no other skills to take up any other job, this leaves them with no alternate options other than staying back in their home.

A fisherman family's socio-economic condition depends to a large extent on the economic functions it executes, i.e. fishing and alternate occupations in the lean period. Economists for long have considered the family as a single economic unit or the agent of development with the argument that members of the family hold similar concerns and interests. The problem with this approach however lies in the fact that in contemporary times, intra-family dynamics have severely influenced the behaviuor of families; and more than the economic aspects, it is the psycho-social issues which are impacting the family income. With pressure on day-to-day livelihood, there seems to be a disruption in the family structure of fishers, as the traditional family arrangement in the fishing household has not been able to match the size and flexibility of contemporary business.

7. Marriage and Gender Related Responsibilities

Gender disparity is one of the strongest factors contributing to large-scale inequities in human development across the globe. The discrimination and inequality have become even worse with the tenacity of poverty, reflecting that when women and men do not have equivalent rights to take part in decision-making, there are direct economic and social implications. These not only have an impact on women but also have consequences for their children, communities and the larger society. The role of women in the milieu of advancement is substantial, not only because they signify nearly half of the population in the countries, but also congruently bear the burden of daily struggle for survival in most of the developing or underdeveloped nations. In the past several years, women's participation in the labour market has been studied closely from the point of view of gender-based occupation and its impact. With globalization and privatization in the 1980s, women's participation in both labour–intensive and technical sectors has experienced a radical shift in India.

In the fisheries sector, widespread poverty is among the most significant issues, particularly amongst traditional fishers contingent on open water bodies and trying to make a living from the small catches of over-exploited waters. Like the other informal sectors, despite their presence at all levels and in all areas of the fishing sector, the role of women in Indian fisheries has stayed largely unnoticed and less deliberated. Though their contribution to the sector has remained important, yet the support has hardly been acknowledged and recognized. In this respect, it is important to highlight Kusabe and Kelkar (2001):

> Studies of women in aquaculture, especially in Asia where aquaculture has long been practised, show that women's labour contribution is often greater than men's although there is almost a complete absence of macro-level aquaculture-related gender disaggregated data. Since they are undertaken by women, they are often viewed as requiring little skill. This is a major gender issue in aquaculture and leads to women's work being undervalued and poorly rewarded. This means they have little control of resources and have minimal access to knowledge, training and new technologies. At the same time, women carry a huge burden of work.

The traditional marriages in the fishing community follows a principle, where the husband takes on the primary role of the wage earner and the wife that of a homemaker. As marital status is one of the vital determinants of social status, it was taken up during the study to determine the social status of the respondents. The marital distribution shows that 86.4 per cent of the respondents are married and 13.6 per cent are unmarried and none of them are widowed or divorced. Fishermen get additional support from their spouse in terms of generating extra income. And this contribution of women is especially vital during the lean seasons, i.e. the off-seasons for fishing which demands for exploring other wage-earning avenues.

However, the women are not allowed to go for fishing in the lake due to the superstitious belief that they are inauspicious, yet they are involved in various other activities related to the fishery business. During the FGD in the field it was confirmed that there is a common religious and traditional superstition of the fishermen that if women go for fishing, there is possibility of inadequate fishing or mishaps like boat crashes. During the one-to-one interaction, the fishermen observed that women have very little contribution to income generation; they are primarily involved only in household work or small tasks like making dry fish. On the contrary, during group interactions with the fishing women, some of the smart and outspoken females informed us that the women play a vital role in negotiation and fixing the selling rate for the dry fish marketing. They even informed us that the selling of fish, preserved dry salted fish eggs and small fish which are sold in several parts of Odisha, and West Bengal and in international markets is handled exclusively by women which fetch a substantial amount of money. However, they also stated that these activities are usually undertaken by relatively more literate and women having newer market exposure. The women through the conversations affirmed the significant contribution and key role played by them giving shape to the fishing family's income, which is underestimated by men. Many of the responsibilities played out by women on the domestic front constitute what feminists rightly recognize as 'unpaid work' (Acharya, 2005). The interesting point is that even where women control the finances their roles are undermined, demonstrating that gender hierarchies pervade both the 'paid' and 'unpaid' sector of fisheries. Mr. Padhiari as I have mentioned before, highlighted family support to be the most important for survival in this fragile environment.

> We have some farm land to cultivate here and we produce high priced vegetables in our land. Production is good, and qualities of vegetables are equally good. My wife helps in our farmland in terms of experimenting. There is no false sense of prestige about the work of women in our community. Vegetable cropping always needs constant care, which my wife does much better than me. Also, my long hours of duty for fishing never allow me to spend time for our farm-land.

In the lean fishing months of summer, when the lake shrinks, the dry bed is sometimes used for cultivating paddy or millet, which is usually done exclusively by fisherwomen. Between February and May, when tiger shrimp fingerlings are found close to the lake banks, the women collect the shrimps with mosquito nets. They catch the baby shrimps, which fetch them an appreciable amount of money per day, which in turn helps them contribute in strengthening the economic situation of their families. In this context, it is important to identify that women support a crucial component of marketing and selling in the fishery livelihoods of fisher households. However, as the role and contribution of women in Chilika to the intra household livelihood cycle as well as to the larger fishery sector is undervalued and at best termed as 'casual'. Thus the interventions required to support them in enhancing the end economic output of the fishing livelihood also remain largely ignored. Further, as discussed in the subsequent chapter, mechanization in the fishery sector has generated a demand to have more skilled labour, where women would find it even more difficult to locate themselves.

8. Income Earning and the Sharing of Labour

The economic situation of households in terms of income is based on the nature of employment and earnings from various kinds of engagement or occupation. In the informal sectors, income and earnings to a large extent depend on the extent of labour and investment. The fisheries industry includes multiple activities that enhance income generation through culturing, processing, preserving, storing, transporting, marketing or selling fish or fish products. The earning from fishing business also depends on the number of family members engaged in fishing as well as the nature of the property.

In fishing production activities are largely carried out by traditional fishing communities, whereas indirect fishery-related activities are mostly looked after by people of non-fishing castes. We therefore made an attempt to understand different fishing activities, of the fishermen. Based on the study, the respondents can be divided into four broad categories; thus 42.5 per cent fishermen are involved in catching fish, while 54 per cent are involved in both catching and selling fish. Only 3 per cent fishermen are involved in selling and 2 per cent involved in networking, including facilitating the process, in terms of catching, selling, and processing. This implies that the fishermen who catch fish are also involved in direct or indirect marketing of the fish. It is important to notice that while fishing from lake takes considerable time, selling the same is also not an easy job, unless there is an existing system in place. This proves that those who lack the capacity to invest continue with simple labour oriented methods, despite the existence of more efficient means of production and processing.

The monthly family income of the fishermen in the Chilika region can be computed as the total earnings flowing into a household from fishing and other sources. Despite the difficulty of obtaining income data for self-employed fishermen households, a framework was generated to estimate monthly incomes with a fair degree of accuracy. Out of 450 respondents studied, the income of 3.3 per cent respondents was below Rs. 2,000, that of 24.9 per cent was from Rs. 2,000 to less than Rs. 4,000, that of 52.9 per cent ranged from Rs. 4,000 to less than Rs. 6,000, while among the remaining households, 9.3 per cent of them earned between Rs. 6,000 and less than Rs. 8,000, and 6.0 per cent households had an income range of Rs. 8,000 to less than Rs. 10,000. Finally, only 3.6 per cent of them reported to earn a family income of Rs. 10,000 and above. The incomes are based on fixed shares of the catch. Low income combined with the number of dependents contributed to increased poverty.

Income earning is best understood against the backdrop of poverty. Poverty in simple terms can be defined as a situation, where an individual or a household lacks a certain amount of material possessions or money. Poverty in India has always been debated, deliberated and decided with immense fascination and curiosity, though the identification of real poor has always persisted as dubious and has remained as a substance of disquiet for the government. The two major questions, which have called for attention from almost all classes and sectors of population are, "who is poor" and "why is there poverty"? Historically, economists in the country have looked at poverty from monetary aspects, social scientists from deprivation and marginalization aspects, and others have combined both. Many philosophers, including the economists also claim that "people are poor, because they are born poor". Therefore, the poverty reduction programmes in India have received key focus since independence. Irrespective of arguments, counterarguments and support from government, the poor have remained poor in India and it has even become an intergenerational cycle for a considerable part of the population. People who are poor also get into a vicious cycle of deprivations, leading to a highly unequal society. In the recent past, the Government of India data on poverty, based on new methodologies of estimation has resulted in the upward revision of poverty. For example, the all India estimates for the headcount ratio of poverty was revised from 27.5 per cent to 37.2 per cent as per the 'Report of the Expert Group to Review Methodology for Estimation of Poverty', Planning Commission, Government of India, 2009. Reardon and Vosti (1995) highlight that poverty in the context of fishery resources came onto the agenda of development practitioners including an agency like UNFAO as a result of the indications that there are links, although complex, between poverty and the degradation of natural resources. Several studies reflect upon the nature, magnitude, effects of poverty in the fishery sector (Macfadyen and Corcoran, 2002; Béné, 2003). There are also studies which have highlighted the vulnerability of fishers in the context of the open access nature of fishery resources (Allison and Ellis, 2001). Béné, (2003) further claims that it is no longer worthwhile to categorize whole communities as "poor", as different strata within the society respond differently to policies and mechanisms to reduce poverty.

We took two broad parameters, i.e. Below Poverty Line (BPL) and Above Poverty Line (APL) to assess the poverty level of the respondents. 'Below Poverty Line' is an economic benchmark and poverty threshold used by the Government of India to indicate economic disadvantage and to identify individuals and households in need of government assistance and aid; while the remaining households are categorized as APL. With the methodological complexities and considering the lack of evidences in form of documentation proof, we used two primary methods to assess the situation. Considering the family size, we used the MDG based poverty rate of less than $1.25 a day per person, where no document could clearly provide us the alternative evidence. The study reflects that most of the respondents' families, i.e. 71 per cent are BPL families and the remaining 29 per cent are APL families. The main findings from the survey prove that household incomes were under reported in fishermen villages. This is the reason for a clear discrepancy between the monthly income and BPL categorization. Most of the fishermen as per the outcomes of this research are below the poverty line.

The findings also highlight the fact that there are limited alternative productive work opportunities available in the area, which can become a major concern for the fishers in years to come. The ever-increasing fishers' population faced with limited alternate income-generating activities can lead to large-scale unemployment in future. The educated lot also has a concern of quality of employment and earnings as in the last few years, they are inclined towards more service-oriented work. Fishers' occupational and geographical mobility in case of Chilika is thus constrained and it is important to bring a balance between unobserved heterogeneity, differential pay compensation and the inadequacies in the fishery market to address pay variance. For example, the caste-based fishing techniques ensure that with almost similar skills sets, one group secures the larger share of the catch, while the other group is usually unable to match the harvest and generates lower resources.

9. Health

Since its formulation, within medical anthropology, the narratives outlining the concept of illness has developed in two different directions. The first is greater refinement of the narratological qualities of conveying the experience of illness and the second is greater attention to the economic and political conditions within which illness experiences are formed (Das, 2015). Clara Han (2013) provides a rather fascinating argument of the unravelling of care and violence in neo-liberal Chile, where she focused on the fact that there can be acute moments including illness that can affect the lives of the families, which often cope through elusive forms of support provided by friends, or neighbours but at other times goes far beyond what neighbours or friends could offer by way of help. Illness therefore reflects that daily life might be any or all of the following. Recent findings on healthcare behaviour confirms that the poor visit medical practitioners more frequently than the rich (Banerjee and Duflo, 2011). This is particularly a consequence of the fact that a wide variety of the practitioners are available in the markets that function in the vicinity of neighbourhoods in which the poor live. There seems to be an increasing demand for receiving treatment for ailments that could have been treated at home earlier. People mostly try and use pharmacies and practitioners to get diagnosis and treatment. Although a detailed analysis of the differences by locality and income is not possible, we can argue that the rich and poor tend to have different types of diseases.

Long hours of tenacious work, extreme climatic conditions and sometime working with equipment contribute to health hazards within fishermen. Despite a common understanding of the fact that the hazards and health-related risks associated with fishing, there has been minimal research conducted in this field. Despite the fact that there has been progress within the fishery industry in terms of safety, little emphasis has been provided on the relationship between health, the environment and performance at work. Kula Behera from Ganjam narrated that,

> Because practitioners working in these kinds of locations are not highly competent and cannot do correct diagnosis they usually treat the illness on the basis of symptoms alone. The matter of the fact is even fever like malaria common in this area can take up to three months to be diagnosed by the local practitioners.

These practitioners are frequently the point of first contact for the patient, despite public health campaigns and check lists of symptoms that are distributed by the government (Reddy et al. 2014). During our discussion we could see that the burden of chronic diseases was more among the older fishermen, partly because of the fact that due to availability of basic healthcare services the diseases like diabetes and hypertensions could be recognized early. The personal interactions with some of the fishermen also brought out a different story as against the narratives mentioned by the local medical officer, who claimed that the fishermen maintain a healthy life and changing climate has hardly impacted their lives. On the contrary, during the field research, on asking specific questions, some of the fishermen showed their feet, afflicted by a serious skin disease. They informed us that these were commonly found in the area and mostly occur during the rainy season. Though they obtain medicines from the local health sub-centre, as per their accounts, there has been no specific extension of government health support in terms of steps to be taken to control them.

We also observed that an important correlate of health is the amount of time the fisher folk spend at their work, and the amount of free time that is available to them. Fisheries being a self-employed job, no such working hours are fixed. In Chilika, most of the time the amount of fish catch depends on the number of hours someone spends in the lake. This also varies depending on the season. Sometimes, the fishermen are forced to spend more hours in the lake to catch the minimum amount for their survival. On doing a basic calculation, it was found that for 35 per cent of fishermen there were no fixed hours for fishing and it depended on the catch. However, 32 per cent fishermen spent more than 8 hours inside the lake for fishing while 17 per cent of them answered spending around 4 to 8 hours and 15 per cent fishermen, up to four hours a day. While the fishermen go to the sea for fishing, they take a longer time to return, which affects both their mental and physical health. As there is no clear-cut guideline for an informal society like the fishing industry, it is difficult to establish a set of strategies to regulate the hours of fishing for the individual fishermen and this aspect needs further in-depth research from the well-being perspective.

It would also be critical to highlight the fact that the lagoon attracts over 10 lakh birds, mostly migratory ones during the winter season, making the whole area susceptible to diseases like bird flu. During the recent past (January 2012) the state's first bird flu virus (H5N1) was detected from this area. During the rainy season, the fishermen of Chilika also become prone to other diseases like malaria, measles, etc. Any outbreak can seriously affect the livelihood of the communities for a period and proves true for the fishermen's community, as they solely depend on catching fish. We also interacted with the fishermen to ascertain, whether the government takes enough measures, before the outbreaks or epidemics to understand the fishermen's vulnerability to the outbreak and also its impact on their livelihoods. It is important to point out here that the nature of long working hours within the lake and sea also makes it difficult to implement health-related programmes which can potentially help the fishermen. Furthermore, the low level of education among the fishermen characterizes another cause of neglect for precautionary measures. However, it is important to have a specific health and safety strategy for adapting to the local conditions and keeping in mind the larger macro environment, so that it becomes more effective.

Based on the findings, I argue that while poverty or low income may not directly result in poor health, the overall socio-economic factors like age, education, housing and drinking water facilities eventually impact the overall health system of the fishers in the region. The interactions with the fishermen point to threats of consumption deficits and long-term human development apprehensions that cannot be ruled out. Another area, which will remain as a major concern is lack of formal insurance and risk protection mechanisms extending the much-required coverage against health shocks and illnesses. These mechanisms can be costly and hence, difficult to be financed solely out of pocket by the fishers and therefore state-specific insurance schemes need to be explored.

10. Housing and Hygiene

A congenial environment at home ensures the well-being and development of family members to a large extent. A good house, whether owned or rented, is largely the outcome of a decent parental source of income (Vandivere et al., 2006). Poor housing could exhort certain environmental conditions causing and aggravating the symptoms of asthma, the most common chronic disease among children (Breysse et al., 2004). In addition to asthma and respiratory distress, poor physical quality of housing may lead to accidents, injury or even death. All these hazards are preventable and treatable, but if not addressed, lead to significant healthcare costs for the household (Vandivere et al., 2006), and impacts human capital formation. House ownership gives a sense of stability and security and to a large extent decides the health and hygiene status of a family. In such circumstances we focused on the status of housing among the fishers in Chilika.

During the study we looked into the number of fishermen that lived in rented shared houses, rented independent houses and or their own houses as the parameters. Out of the total respondents 2.2 per cent lived in rented shared houses, 0.7 per cent in rented independent houses and 97.1 per cent live in their own houses. While owning or renting a house is generally a matter of choice and economic capacity, it is one that is heavily constrained for the fishermen of Chilika. Most of the houses that we visited were thatched houses, and many were in a dilapidated condition. In fact the dwellings in which most of the fishers live do not enable them to enjoy the benefits of enjoying their free time. The fisher-specific scheme, though provides support for building houses, do not have any provision for support on maintenance help in making repair work, which is a massive task. This gains special significance in light of the fact that most of the houses are built in ecologically fragile zones and are prone to easy and recurrent destruction by rough weather and strong winds. Physical observations of the household reflected that congestion in living spaces is high compared to any other areas and a few of the fishers narrated that over the years it has even increased. Considering the relative availability of open spaces in rural areas, such congestions in living spaces indicate poverty and subsequent inability to construct additional spaces.

Water sourced from taps, wells, tube/bore wells, hand pumps are in general considered as safe sources for drinking purposes. Water sourced from a spring, river, canal, tank and lake are not considered as safe sources. Unsafe sources of drinking water have a positive correlation with acute illness of diarrhoea and dysentery among rural population, especially among the children. Improved access to safe and clean water and sanitation facilities also can potentially affect issues specific to education, health, environment as well as poverty among fishermen. However, most of the time the fishermen's households lack clean drinking water, which can potentially affect their livelihood, as it can bring and further spread disease. Almost 57 per cent respondents felt that the government has not taken any such measure for supplying clean and safe water in the locality, whereas the rest were positive in terms of the government's response.

While trying to assess the drinking water facilities in the village, the Panchayati Raj Representative from Balia village provided a positive picture informing us that being close to the lake, the villagers never faced any problem for drinking water. However, during interactions with the local women, some of them insisted I drink water from one of the dug wells and while tasting the water, I realized that the taste was not good, although it was difficult to estimate the actual quality of the water. The women informed us that because of the lake, they get water from a low level, but the quality of water is questionable. Two of the prominent authorities on water issues in India, M.D. Kumar and Tushar Shah, in co-authoring a paper on water issues, have highlighted that some of the major pollutants in ground water include fluoride, salinity, iron, nitrate, arsenic and heavy metals (Kumar and Shah, 2006). They further warn that unlike salinity and alkalinity, fluoride cannot be detected through "tasting." Water testing kits become essential in such circumstances, which were unfortunately beyond our reach. The reasons and causes of ground water pollution await greater attention and remains an endemic issue in the region. Also for years together, the absolute practice of fetching water from ponds and open wells has been common in Chilika. Most of the time the fishermen face the health hazards of hookworm infection, diarrhoea and cholera by consuming poor quality of water. The ground water available in the shore area of the Chilika Lake is generally brackish in nature and is not suitable for drinking. However, there is an obvious lacuna in the policy front wherein there is a need to revisit and design interventions keeping the contextually relevant issues related to housing facilities of the Chilika population central to the policymaking.

Conclusion

Demographic features are the primary variables to analyse socio-economic conditions of any group. The socio-economic approaches have traditionally focused on the physical resource base to explain inequality without really considering other components of the structural detriments faced by the commons. How households are socially embedded in caste, class and other categories that impact their livelihoods is often bypassed (Newton, 2007). Further comprehensive research on the socio-economic position of fishers in Chilika is limited. Since collecting all the available statistics was impractical and impossible, we focused only on key quantitative demographic and other related indicators; though to our surprise the discussions with respondents individually and in groups, astonished us with several important aspects about the fishing community. Discussions on the principal ways fishers make their living, and the share of each activity in the priority area or the sector's overall economy helped us in understanding the economic aspects of the fishers' lives. Similarly, the complex relation between variables like caste, gender and health and hygiene in the context of fishery livelihood reflected on the social aspects including how fishers perceive, act and negotiate their well-being in the context of decreasing fishery resources. This enriched us with the fact that with increase in urbanization and mechanization, the fishers are dealing with a highly volatile situation, where adaptability becomes crucial. The overall findings reflect that irrespective of the contribution of fishers in the areas of food and income, their economic conditions have not improved with reference to the past. The effect of government welfare schemes linked to fishery and fishers' well-being have hardly made any significant impact due to the multiple-use common pool nature of the lake, which we will discuss in the subsequent chapter. On the other hand, the fishers' perception of poverty as deficiency of resources, rather than absence of entitlements, further brings the logic of violation of rights, which need further research.

NOTES

1. Jappelli (1999) and Kapteyn et al. (2005) explored why different cohorts do not generate the same amounts of capital. Their study concludes that output growth is the primary element of variances in wealth across cohorts. Efficiency growth produces alterations in permanent incomes across cohorts, which feeds into the wealth accrual of individuals belonging to different age groups.
2. Both Dhaudi and Baza are kinds of bamboo made trap fishing box. While Dhaudi is larger in size, Baza is smaller.

REFERENCES

1. Davies, J. and Shorrocks, A.F. (2000), The Distribution of Wealth. In James Davies and Anthony F. Shorrocks (eds.), *Handbook of Income Distribution*. Vol. 1. Elsevier.
2. Wisner, B., Blaikie, P., Cannon, T. and Davis, I. (2003), *At Risk: Natural Hazards, People's Vulnerability and Disasters*. 2nd Edition. London: Routledge.
3. Jappelli (1999), *The Age-Wealth Profile and The Life-Cycle Hypothesis: A Cohort Analysis with a Time Series of Cross-Sections of Italian Households*. Working Paper No. 14, CSEF, Università di Salerno (Italy), and CEPR http://citeseerx.ist.psu.edu/viewdoc/download? doi=10.1.1.40.7071&rep=rep1&type=pdf
4. Kapteyn, A., Alessie, R. and Lusardi, A. (2005), Explaining the Wealth Holdings of Different Cohorts: Productivity Growth and Social Security. *European Economic Review*, 49: 1361-1391 http://pages.stern.nyu.edu/~dbackus/BCH/saving/KapteynAlessiaLusardi_ wealth_EER_05.pdf
5. Allison, E.H. and Ellis, F. (2001), The Livelihoods Approach and Management of Small-Scale Fisheries. *Marine Policy* 25, 377-388.
6. Pollnac, R.B., Pomeroy, R. and Harkes, I.H.T. (2001), Fishery Policy and Job Satisfaction in Three Southeast Asian Fisheries. *Ocean and Coastal Management* 44: 531-544.
7. Madan, T. (1993), Whither Indian Secularism? *Modern Asian Studies, 27*(3), 667-697. Retrieved from http://www.jstor.org/stable/312965.

8. Madan, T.N., ed. (2004), *India's Religions: Perspectives From Sociology and History*. Delhi: Oxford University Press.
9. Gadgil, M. and Vartak, V.D. (1975), *Sacred Groves of India, A Plea for Continued Conservation*.
10. Rangarajan, Mahesh and Saberwal, Satish (2003), *Battles Over Nature: Science and the Politics of Conservation*. New Delhi: Sangam Books, p. 412.
11. Jalai, A. (2009), *Forest of Tigers: People, Politics and Environment in the Sundarbans.* New Delhi: Routledge.
12. Ninian, A. (2008), India's Untouchables: The Dalits. *Contemporary Review*, Vol. 290 (May), pp. 186-192.
13. Thekaekara, M.M. (2005), "The View from the Top." *New Internationalist Magazine*, Issue 380, July.
14. Gadgil, M. (1999), Promoting Adaptive Participatory Management. In P.J. Schei, O.T. Sandlund and R. Strand (eds.), *Proceedings of the Norway/UN Conference on the Ecosystem Approach for Sustainable Use of Biological Diversity*. September, Trondheim.
15. Allison, H.E. (2003), Linking National Fisheries Policy to Livelihoods on the Shores of Lake Kyoga, Uganda. LADDER (Livelihoods and Diversification Directions Explored Research) Working Paper No. 9, University of East Anglia: Overseas Development Group.
16. Meinert, L. (2003), Sweet and Bitter Places: The Politics of School Children's Orientation in Rural Uganda. In: K.F. Olwig and E. Gulløv (eds.), *Children's Places: Cross Cultural Perspectives*. London: Routledge, pp. 179-196.
17. Ansell, N. (2008), Substituting for Families? Schools and Social Reproduction in AIDS-affected Lesotho. *Antipode 40*, 802-24
18. Lewin, Keith M. (2007), *Improving Access, Equity and Transitions in Education: Creating a Research Agenda.* Project Report. Consortium for Research on Educational Access, Transitions and Equity (CREATE), Falmer, UK.
19. Maddox, B. (2006), Literacy in Fishing Communities. Rome: FAO. Available at http:// www.fao.org/newsroom/en/news/2006/1000290/index.html (accessed October 11, 2009).
20. Sen, A. (1992), *Inequality Reexamined*. Cambridge: Harvard University Press.

21. Sen, A. (1999), *Development as Freedom*. Cambridge: Harvard University Press.
22. Nettle, D. (2000), Linguistic Fragmentation and the Wealth of Nations: The Fishman-Pool Hypothesis Re-examined. *Economic Development and Cultural Change 48*, 335-348.
23. Belle, D. (1980), *Low-income Depressed Mothers: Parent Intervention*. Cambridge, MA: Harvard University Press.
24. Kusakabe, Kyoko and Govind Kelkar (eds.) (2001), Gender Concerns in Aquaculture in Southeast Asia, Gender Studies Monograph 12. *Gender and Development Studies*, Thailand: Asian Institute of Technology.
25. Acharya, M. (2005), Accounting for Women's Work in Censuses in Developing Countries–Nepal Experience. Paper Presented at Global Conference on "Unpaid Work and the Economy: Gender, Poverty, and the Millennium Development Goals", October 1-3, New York.
26. Reardon, T. and Vosti, S.A. (1995), Links Between Rural Poverty and the Environment in Developing Countries: Asset Categories and Investment in Poverty. *World Development 23*, 1495-1506.
27. Macfadyen G. and Corcoran, E. (2002), Literature Review of Studies on Poverty in Fishing Communities and of Lessons Learned in Using the Sustainable Livelihoods Approach in Poverty Alleviation Strategies and Projects. In: FAO Fisheries Circular No. 979. Rome: Food and Agriculture Organization of the United Nations.
28. Béné, C. (2003), When Fishery Rhymes with Poverty: A First Step Beyond the Old Paradigm on Poverty in Small-Scale Fisheries. *World Development 31*, 949-975.
29. Allison, E.H. and Ellis, F. (2001), The Livelihoods Approach and Management of Small-Scale Fisheries. *Marine Policy 25*, 377-388.
30. Das, Veena (2015), *Affliction: Health, Disease, Poverty*. New York: Fordham University Press.
31. Clara Han (2013), "A Long-Term Occupation: Police and the Figures of the Stranger", and "More Eyes, Different Eyes: A Response to Didier Fassin". *Social Anthropology* 21(3): 378-384; 387-388, August.

32. Banerjee, Abhijit V. and Duflo, Ester (2011), Poor Economics: A Radical Rethinking of the Way to Fight Global Poverty. *Population and Development Review*, 37: 796–797. doi:10.1111/j.1728-4457.2011.00462.x
33. Reddy, Sreerama, C.T., Qin, Z.Z., Satyanarayana, S., Subbaraman, R., and Pai, M. (2014). Delays in Diagnosis and Treatment of Pulmonary Tuberculosis in India: A Systematic Review. *The International Journal of Tuberculosis and Lung Disease : The Official Journal of the International Union Against Tuberculosis and Lung Disease*, 18(3), 255–266. http://doi.org/10.5588/ijtld.13.0585.
34. Vandivere, B.S., Hair, E.C., Theokas, C., Cleveland, K., Mcnamara, M., Atienza, A., and Vandivere, S. (2006), *How Housing Affects Child Wellbeing.*
35. Breysse, P., Farr, N., Galke, W., Lanphear, B., Morley, R., and Bergofsky, L. (2004), The Relationship Between Housing and Health: Children at Risk. *Environmental Health Perspectives*, 112(15).
36. Kumar, V., Shah, D., Venkatesan, R. (2006), Managing Retailer Profitability: One Customer at a Time. *Journal of Retailing* 82(4), 277-294.
37. Newton, J. (2007), Structures, Regimes, and Well-being. WeD Working Paper No. 30, http://www.welldev.org.uk/research/working.htm#wed30.

4

The Baits and Bytes of Fishing in Troubled Waters

In the previous chapter we explored the living conditions of the fishers within a socio-economic framework, paying close attention to various factors such as age, caste and traditional patterns of resource use, religious backgrounds, earning capacities, and gender dynamics among the members of the larger community. Our central focus has been the argument that livelihoods are impacted by various factors including caste and class that individual households are embedded in, and it is this connection that is often overlooked (Newton, 2007). The fisher vulnerability I will be subsequently discussing is strongly linked to property rights in the context of Chilika. It is argued that in the case of open access and unregulated common property, individuals do not get appropriate incentives to act in a socially efficient way. Guided by Demsetz's path-breaking article on "Towards a Theory of Property Rights" many continue to argue that private property is the most efficient way to internalize the externalities that arise when access is unregulated. It also suggests that private property rights will spontaneously emerge to increase efficiency (Demsetz, 1967). However, failures of market and state institutions as a cause of unsustainable natural resource management, have also been discussed in recent literature. The central thesis of this book revolves around the issue of common property and how each of the aspects discussed need careful discernment in the context of regulation to sustain and promote traditional fishers' livelihood in Chilika.

This chapter thus sheds light on conflicts emerging from user rights of commons. 'User rights of commons' basically connote claiming of greater rights by the locals over the common property as against the outsiders' entry into the area for fishing; and the nature of protection given to the former by the state. User rights of commons centres around the concept of common property resources (CPRs). The fisher community living around the lake while having greater usufruct rights on the waters for fishing, are in conflict with the 'outsiders' on issues specific to undue and unsustainable use of resources. Similarly, the local fishers assail the state machinery for not protecting them against the outsiders' use of Chilika resources and decry the absence of clearly defined guidelines. While the people's movement for protection of rights of the fisher community is gaining momentum, conflicts are visible at various levels, exacerbated by non-cognizance of government rules and regulations. The central argument of this chapter reflects the conflicting situations that emerge over lack of protection to the locals and absence of clear cut guidelines by the state as to the claim over user rights to the common.

Collective Action in Managing CPRs

Common Property Resources derives its origin from the medieval manorial system, when much of the common lands were wastelands and other resources like water bodies, hills and forests, were community controlled to ensure entry and access to such resources. Resource use arrangements by different groups in case of CPR, confirm how norms and guidelines established by people themselves are important in dealing with sustainability of resources. Throughout history, the need for social regulations of common property resources has been well recognized. Klitzner et al. (1988) argued that although all the norms are not considered to be formal laws within society, they still work to promote a great deal of social control. Resource access in Chilika, as an example, is primarily dependent on community management norms, and how communities perceive their relationship with resources. However, major areas of concern in the context of Chilika include the issues of ownership of the lake, which have their direct bearing on the fishers' existing livelihood structure. The fishing boundaries within the lake, as well as the access restrictions are determined based on community understanding. However, over course of time, increased dependence, along with unsustainable use of common property, has affected these norms. Social theories and ecological arguments highlight the need of having a set of guidelines (whether state or community) to manage the property well. Much of the solution lies in the ability of the commons to collectively identify, manage or avoid the adversities. They often fail to decide on the rules that can afford long-term protection to the commons, unlike institutional norms strengthened by trial and error methods. Evidence from Turkey shows how the fishing cooperatives have been sustained by the state, for management of inshore fisheries, by recognizing their joint survival as juridical persons with the right to sue to safeguard their property (Berkes, 1968).

Fishers follow the access rules and restrict those who intrude on these rules. Such understanding among the fishers helps in avoiding overexploitation and competition for the limited resources. The problem arises when individuals are expected to cooperate with a larger community to achieve individual and collective goals, at the cost of the cooperative short or long-term benefit. Therefore, most of the CPRs deal with an intrinsic cost-benefit analysis at the community level. Analysis of this factor can tell us why people continue to use unsustainable methods of fishing, even while comprehending its long-term socio-economic adverse impacts. The pattern of resource use by communities is a reflection of indigenous norms and rules developed by different fishing groups, and their significant influence on the sustainable ecological management of resources. However, in recent years local resource management models have become vulnerable to the socio-economic and policy changes induced by the state, which is evident from the increasing number of conflicts and violation of rights by traditional fishers in the lake areas. While the size of the lake is shrinking with decrease of the water levels, there are added problems of acute siltation. Anecdotal evidence shows that the numbers of brackish water species are declining, impacting the quality of freshwater, which is simultaneously undergoing a visible change. There are therefore strong pointers to the overall ecological and the natural habitat degradation of the lake, invariably resulting in negative externalities with respect to the socio-economic conditions of the traditional fishing community of this region.

The right of enacting laws or policies in India is divided between the central and the state government. Subject matters including water, fisheries along with preservation, protection and improvement of livestock, fall under the purview of state legislatures, to pass laws and make regulations. Thus, the fishery sector is considered to be a state subject and each state has a full-fledged Fisheries Department. The Department of Animal Husbandry Dairying and Fisheries (DADF) under the Ministry of Agriculture, Government of India, provides additional support of development programmes in different states along with funding for centrally sponsored projects. Central assistance on fisheries is mostly under the heads of marine fishery and management of the central fisheries institutes. The most relevant acts and policies with respect to fisheries in India are the Inland Fisheries Act or the respective leasing and licensing policies of water bodies formulated and implemented by the states. However, the origin of India's Fisheries Act can be traced to the historical Indian Fisheries Act 1897, which prohibits the state from unsafe fishing practices in the open areas. It also imposes restrictions on allowing fishing in a particular area of the water body, in a specific season to be regulated by the government. The other acts which have a substantial impact on the Indian fishery sector are Environment (Protection) Act (1986), which is an umbrella act containing provisions for all environment-related issues affecting fisheries and aquaculture industry in India; the Water (Prevention and Control of Pollution) Act (1974) and the Wild Life Protection Act (1972). The Chilika Bill, which has subsequently been discussed at length, is a state subject and is the responsibility of the state government. The regulations in the fishery sector are intended to largely help in sustainable management of fishery resources and their habitats. In fact, licensing and leasing rules for inland water bodies for capture and culture forms the core area of most inland fisheries policies. Fishing rights in Chilika are governed by several acts, policies and regulations[1]; with individual connotations and varying applicability in defining common property.

In Odisha, one of the most important Acts that govern the marine fisheries is the Orissa Marine Fishing Regulations Act (OMFRA), 1981 by which the state regulates the certification, registration and licensing of fishing vessels and the granting of fishing licenses. The most recent development is the declaration of the Comprehensive Odisha Fishery Policy 2015, with the broader objectives of positioning Odisha as a leader in aquaculture development and fisheries extension for ensuring food security, livelihood, welfare of fishers and employment generation by 2020. With several welfare and business opportunities, this policy has the potential to change the state of the Odisha fisheries sector by 2020. As this is more of a recent development and the results are yet to be seen, I have therefore primarily focused on the elucidation of my research findings.

Common resources, particularly the land and water bodies have considerable environmental importance and play a significant role in delivering the state's obligation for improving environmental well-being. Although common properties are usually owned by the state or even private parties, resources including the open fishing areas are subject to rights over commons, which are primarily based on local traditions and social norms. Keeping in mind the ever increasing concerns surrounding the commons, most of the scholars in the past have focused their analysis on critically examining Hardin's 1968 paper "The Tragedy of the Commons". Even the *Encyclopedia of Political Economy* opines, "the common property 'problem' was originally articulated best by Hardin" (Dragun, 2001). The work of two eminent scholars Elinor Ostrom (1990) in her book *Governing the Commons*, and David Harvey's (2011) paper "The Future of the Commons" have particularly focused their research based on Hardin's theory. Hardin specifically claimed that common pool properties tend to be mismanaged because people are individualistic, self-centred, and profit-oriented. His argument was based on the fact that common property is individualistic, and its ravages are further strengthened by population growth. Based on his pessimistic conclusion that freedom over commons brings ruin to all, Hardin reclaimed privatizing the commons might be unfair, as that "[t]he alternative of the commons is too horrifying to contemplate (Hardin, 1968). The particular aspects which Hardin focused on were the outward incapability of modern societies to deal with environmental issues and the need of collective action of some sort. His solution of public action was framed as one of only two possible alternatives by which pollution of the environmental commons could be dealt with, either ineffectively through private market action or effectively through governmental or public action. Against Hardin's argument, most contemporary scholars have highlighted that sustaining environmental resources is not dependent on a structure of property rights regime. Instead, it is influenced by a well-specified property rights regime; and a congruence of that regime with its ecological and social context (Hanna and Munasinghe, 1995). Bollier (2007), further mentions the usefulness of commons as it helps in identifying issues that affect both types of commons including congestion, overharvesting, pollution, inequalities, other degradation and to propose effective alternatives, e.g. social rules, appropriate property rights and management structures. By highlighting the two types of commons, Bollier was possibly referring to property and information commons. But the idea is not to get into the kinds of commons here, rather, the objective is to provide a larger picture of commons in the context of fishery resources.

Against all these theories, the Ostrom model is more prescribed, broad, balanced, ecological, and has been applied to a far wider range of commons. Commons studied by Ostrom includes not only fishing grounds and other settings where humans aggregate, decide and act, but also factors in physical topics. While considering commons, Ostrom highlights that mutual and cooperative action for common property management will become effective, if well-defined boundaries with set guidelines for conflict resolution, along with effective monitoring systems can be put in place.

Common property in the context of Chilika needs careful explanation as it is often confused with other forms of properties from which it must be separated, as individual property rights have different consequences in the context of resource management. Some confuse common property with not owned property and some with public property. A broad definition of commons is almost impossible to arrive at any point of time, as the collective interests and beliefs that produce "commons" is intricate. Even so, a tentative description that spans different forms of commons is "the collective and local ownership of land, resources, or ideas, held in an often communal manner". The arguments however justify that common property rights are complex and multidimensional, as in this case, the rights have no demarcation or delineation. No one has the right to eliminate others from using them, and all are free to use them at their will. However, in most of the cases these rights are driven by social norms, customs or traditions, sometimes limiting user access to the property and mostly to ensure sustainable management of the property. Therefore, different property rights not only have dissimilar production effects, but also have different allocation consequences. It is therefore important to consider the relationship between equity and distribution of existing traditional institutions, for the management of CPRs, under growing inequality on the supplies of assets and opportunities in the context of fishers in Chilika.

The local fishers of Chilika consider the lake as a common property with the spatial access rights governed by them. However, in the last few decades, increased dependence on the lake along with unsustainable fishing practices have led to state interventions. Conventionally it is acknowledged that common property resources give rise to limited solutions, as user rights on them are not strictly defined. While user rights not being defined is one part of the problem, other parameters such as the size and the heterogeneity of the user groups, pose problems in the cost-benefit calculation. Therefore, either interference of the market or the government, in the form of price mechanism and institutional arrangements, are often suggested as solutions to the problems of the commons. In the most basic premise, the lake reflects that the fishing rights in Chilika do not have a market, where the rights can be sold, bought, leased or exchanged. Rights are acquired or granted to a section of fishers based on their traditional fishing rights and cannot be transferred to other persons via any exchange mechanism.

Determining Fishing Rights and Violation of Community Norms

The fishers' access to the lake depends on its relation with the changing nature of fishery resources, as well as both micro and macro-economic situations, governing the livelihood of traditional fishers, supported by state regulations. For example, the new entrant's entry to fishing is largely dependent on the community's own assessment of property rights and the entrant's own justification for his rights over the resources. Access rights provide some degree of control over resource-use and are considered by communities as a means of regulating resources locally. Thus, those components of access, that local fishing communities are both able to and allowed to control, are the ones that form most of similar ownership claims. As highlighted earlier, fishing rights in Chilika are perceived as a family occupation and the fishers hold a strong opinion about community norms over property rights. Property relations between two or more individuals or a group of individuals, is on one hand, the interests protected by a right and on the other hand, the result of congruent duties. This means having a right also entitles the individual in both law and practice that his/her claims will be respected by those with certain obligations.

The fishing rights of traditional fishing communities in Chilika can be traced to the British period. Though in the intervening period, several government policies have come into effect, it has made no substantial difference in terms of lake ownership; therefore, through this study an attempt is made to observe the fishers' opinion of their fishing rights. The study highlights that a major portion of the respondents (67 per cent) believe that they have inherited their fishing rights from their fathers, 27 per cent from the government and 6 per cent consider that they have inherited such rights from no one. While considering rights, the fishers accord absolute right over the lake waters and do not expect state intervention. With decrease of fish in the lake, entry of multiple players, environmental degradation, and absolute dependence on fishery, the fishers are rendered deeply vulnerable in terms of their livelihood. Within the community, rights to common property resources are unlikely to be either exclusive or transferable; they are often rights of equal access and use (Feeny et al., 1998). However, sometimes institutional failure to safeguard the common property could be attributed to internal reasons, such as the inability of the users to manage themselves, or it could be due to external reasons, for example, an incursion of outsiders (Dove, 1993; Berkes and Folke, 1998). As validated by our study fishing in Chilika is intergenerational; yet the poverty trap is almost inescapable without a constant supply of fishery resources. To address the challenge of sustainable livelihood, there is a need to have structured investigations keeping in mind the "commons", to understand the root cause of poverty in the wider context, as well as within the core of fisher people's livelihoods. Benu Behera from Ganjam said:

> Violations of community rules in the lake area have increased in recent years. We usually intervene and settle conflicts or impose penalties by way of fines. Many a time, restrictions for entry are imposed for a period. This has somehow resulted in creating a fear of violating community restrictions. We have decided that whether a new family or individual who violates rules are allowed to fish; will depend on the village committee meeting, taking into consideration the explanation of the violators, about his/their own rights and claims. If our assessment proves to be adverse, entry is denied. However, in some cases we find fishers to be extremely poor and we leave them with a warning only. We understand together we can save the lake and punishment in all cases may not result in effective management of the lake.

The strong observation about family inheritance of fishing rights further reflects the common property nature of the lake resources. Benu Behera says that the pattern of resource access by traditional fishers has substantially transformed in relation to the past. Easy incomes generated by mechanization and shrimp culture have supported the entry of outsiders, which in turn has disrupted the community norms. Similarly, upper caste, non-fishers keen to enter the fishery occupation, had apparently begun to seek fishing rights from the state. In most cases they encroached the lake either by bribing local people or by providing a substantial amount of money, leading to unauthorized fishing grounds inside the Chilika lagoon. In the past, each fishing ground was identified not by area or boundary but only by nomenclature. The unclear boundary status of fishing grounds enabled non-fishers to encroach on fishing grounds owned by fishers or takeover unauthorized fishing grounds inside Chilika. However, by the time we conducted our study, most of the fishers did comprehend the significance of fishing boundaries.

The common argument that arises from the current situation, points towards the need to have cooperative regulations or what Hardin (1968) described as "mutual coercion jointly agreed upon" and Levi (1988) termed as "quasi-voluntary compliance". However, the significance of independent control over the commons is highlighted by several authors in view of their research on failed common property, where the state undermined the independence and authority of the local unit, that has historically managed the property. Thus, in India, the government "ensured" autonomy over forests, which in reality, had been managed by the indigenous population as common property for generations (Peters, 1983). A clear understanding as to who is eligible to use the commons appears to be necessary and any community which is not serious about this, may not fare well. An eligible user must be a local fisher who would be available to perform his own duty to contribute to the commons, since outsiders who may bring in economic prosperity, may not necessarily be committed to the sustainability of the lake.

Fishing in Chilika for the local fishers is not only a means of livelihood and food security but also tradition. For centuries, the local fishers have evolved a rather intricate system of partitioning the lake area amongst themselves, which is mutually beneficial.

The fisher communities have defined certain mutually agreed upon rules, which include adhering to the fishing boundaries and individual fishing business for their own benefit. However, as seen elsewhere, community norms are violated several times or sacrificed for the need of a smaller group of people from within the community. While it was impossible to investigate violation of community norms individually, the study attempted to find if the fishers have ever been found violating community norms by understanding their experiences. Two-thirds of the respondents (62 per cent) expressed that they respect the community access norms and have never violated them. They further added that they would not encourage other groups also to violate the community norms. However about one-third (38 per cent) of the respondents admitted that they had violated community norms in the past. Most of the respondents, about 78 per cent, reported that the different fishing communities and the groups are aware of their boundaries. Only 22 per cent respondents informed that the fishing communities are not aware of their boundaries. This resilience with regard to community regulations over fishing rights in Chilika is reflected from their relevance among the fishers even today; and points to the efficacy and benefits with respect to the management of fishery resources of the lake. This is precisely the reason for the acceptance of different fishing groups of their boundaries and respect for each other's operational areas. Though the fishing areas of some groups may overlap, but they in general, tend to avoid competition. Therefore, fishers understand that for any property regime, a system of authority is essential, to ensure that the expectations of the right holders are met.

The rules of collective accountability provide fishers within villages and families with positive encouragement to regulate the conduct of their members, so that innocent fishers do not suffer from violations of the guilty. Minor violations including entry into restricted areas or breach of informal boundary rule are considered normal and dealt with caution. Major violations of fishing rights by the traditional fishing community are rare but during our discussions with fishers, a few respondents recapitulated some incidents of major violations in the lake area. One of the stories which was of interest was about a family from one of the fishing villages, who had breached the fishing boundary, claimed to be above the law and no longer wanted to abide by the rules governing the fishers. In this case the whole family suffered a kind of isolation for a specific period, as other fishers socially boycotted them.

Lake Ecosystem and Fishing Rights

Lagoons are coastal systems which play an important role in the reproduction, growth, and protection of many marine fish species. Chilika lagoon, being estuarine in nature has assemblages of marine (southern sector), brackish (central sector) and fresh water (northern sector) conditions which engenders high biodiversity. The cyclical fluctuation in different hydro-physico-chemical parameters facilitates the biological alteration in the ecosystem. Benu Behera along with his wife have seen the changes in the lake eco system. He says:

> The lake was not like this during our childhood. Our parents used to know most of the fishers in the region. Subsequently, with shrimp aquaculture taking a lead role, several outsiders came here. Many of our relatives and friends also ensured their easy entry as they came with raw cash, all black money... now you know... we do not get to see several kinds of fish, which used to be common earlier, several foreign birds have become rare. We do not know whether our children will be equally concerned for the lake ecosystem. The elder one says those who earn more should be held responsible. Why should we be held responsible? But we make him understand the lake is our mother and for our mother there cannot be different responsibility.

For managing the lake eco system, Benu's statement holds significance. Fishing activities can be affected by the destruction of critical habitats where fish species breed or spend part of their lives. Chilika Lake is a hot-spot of biodiversity and shelters several endangered species listed in the IUCN red list of threatened species; however, the species that call the lake their home have suffered a huge loss and destruction to their habitat. Fishes in the lake generally breed during the rainy season. The loss of depth has caused a decline in the availability of deeper shelter zones as breeding grounds, leading to a serious decline in fish catches from the lake. Our study further reflects that more than half of the respondents, i.e. 65.11 per cent would not enter some parts of the lake, especially areas where fish breed during a specific time. They had a clear understanding that use of such areas would reduce their overall catch. However, some of the fishers (35 per cent) were found to be less concerned about this and preferred open access and fewer restrictions. During the survey, fishers from all areas around the lagoon admitted restriction of certain nets in the lake, details of which are discussed in the previous chapter. The respondents from all sectors of the lagoon insisted on eviction of un-authorized encroachments for gheris (*enclosures*[2]) and restrictions on catching of the juveniles of economically important species. This creates a critical niche for substantiating the fishers' indigenous knowledge to identify the general breeding pattern of fishery resources; as most fish species come to the lake from the sea for breeding and feeding. Unless the fishers mobilize themselves to support the ban or restriction of fishing activities during the breeding time, it may not benefit them in the long run. Also the state's support during this period would be crucial to protect the fishers' livelihood.

Figure 4.1: Shrimp gheri in Chilika Lake

Based on the findings, it is suggested that the fishers should be educated to avoid catching brood fishes, especially mullets, as they migrate to the sea for breeding. Further, restrictions should be placed for fishing and entry of boats in certain distinct areas like Nalabanda Island and Rambha. The former serves as the breeding ground of many fish species and shelters migratory birds and the deep waters of the latter serves as spawning areas for fish. The Alimi nets and Box nets used by the fishers in the new mouth of the lake at Satpada, should be strictly prohibited; and some area in the lagoon, should be demarcated to allow fish to spawn, without disturbing the spawning area of fish.

Governing the Lake: The Chilika Bill

Chilika was primarily beyond any state regulations till 1991, although both fishermen and non-fishermen practised an illegal regulation culture. Subsequently in 1991, the lease policy was laid down by the state government, dividing the whole of the lake area into 'Capture' and 'Culture' and allowed the non-fishermen of the locality to involve themselves in 'Shrimp Culture'. The lease policy also led to the initiation of the Chilika Aquatic Farms Limited (CAFL)–a joint venture between the Government of Odisha and the Tata Company which ultimately steered the famous Chilika Bachao Andolan (Save Chilika Movement) against the Tata's project and other illegal shrimp cultivators. In 1993, the Odisha High Court gave a ruling, affirming the rights of traditional fisher folks in Chilika, banning shrimp culture and directing the state government to demolish all shrimp gheries which were illegal. This also ensured that the Tatas moved out from the lake. In 1994, the government revised the 1991 lease policy with an intention to redefine 'capture' and 'culture' sources more meaningfully by defining the role of the Fishery Department, in fishery management of the lake. However, Pattnaik (2007) claims that the revised lease proposal did not make any significant change over the earlier policy, and it was just a window dressing keeping the High Court decision in view. In 1995, the Alagarswamy Report came out with recommendations to save the culture source. The report specifically recommended for withdrawal of aquatic culture farms, prohibited commercial shrimp culture, and suggested renovation of the lake mouth for effective water discharge from both sides. Another important fact pointed out by the Alagarswamy Report was the conversion of paddy fields to shrimp farms and how paddy lands along the fringe of the Chilika were being lost to shrimp farming (AIR, 1997).

The year 1996 is a historical year for Chilika, when the Supreme Court of India taking cognizance of several reports by NEERI, Central Board for Prevention and Control of Water Pollution, and the Alagarswamy Report delivered a significant judgment against aquaculture in the lake. The Court held that intensified shrimp farming culture by modern methods disrupts constitutional provisions and central acts, especially the Environment Protection Act and therefore should not be permitted to operate (Supreme Court case finder, 2000). In 1997, the Odisha legislative assembly constituted a sub-committee to look at shrimp culture in Chilika Lake. The committee still allowed the practice of leasing out some portions of Chilika Lake for shrimp culture.

In the previous sections, it is highlighted that, fishers in Chilika have experienced several bloody fights in the past and many have also lost their lives. One of the most tragic incidents happened in May 1999 at Sorana Village, when the traditional fishers started demolishing large numbers of the shrimp gheris with their country boats. With police interventions and the arrest of the fishers' leaders, the incident took a violent turn with stone pelting; in the police retaliatory firing, three fishermen were killed and injuring several others. This happened in spite of the fact that the fishers had given an ultimatum of 48 hours to the administration for clearing the gheris; and police did nothing in terms of taking any precautionary measures. Subsequently, this led to the constitution of the P.K. Tripathy Commission to enquire into the entire incident and probe the police firing in 2000. In the meantime with pressure from all segments of the population, the government put off lease for shrimp culture in the lake.

Figure 4.2: Shrimp gheri with net enclosure in Chilika Lake

The Government of Odisha passed the Fishing in Chilika (Regulation) Bill in 2002 which protected the traditional rights of the fishers over Chilika but accorded 30 per cent of fishing rights to non-fishers. The Jan Adalat, an association formed by the traditional fishers vehemently opposed and asked for its withdrawal. They considered 30 per cent to be a large mass that could severely affect the overall livelihood of fishers. In fact, while the 30 per cent share may seem like a tilt in favour of traditional fishers, the clause, in fact, virtually sanctions illegal encroachment of the lake's waters by the shrimp lobby and gives non-fisher groups a legal position in the whole fight.

The traditional fishers unanimously made the Odisha Government retreat in tabling the Chilika Bill in the winter session of the Odisha Assembly of 2004, when they came out in protest and held siege of Bhubaneshwar. The "Chilika Matsyajibi Mahasangha", led the mass movement in which massive demonstrations, mass dharnas (protests) and blockade of rail and roads were organized. This unprecedented mass movement, like a huge upsurge, created pressure on the opposition as well as on the ruling government and consequently, the entire opposition in the State Assembly opposed the Bill when it came for discussion. Surprisingly, some of the ruling party MLAs also opposed the Bill, threatened by the mass upsurge. Meanwhile, the Tripathy Commission Report released in 2007, justified the police firings in Sorana which was also vehemently objected to by the Chilika Matsyajibi Mahasangha. Subsequently, in 2008, the Mahasangha threatened to stop the vehicular movement on the roads of the Odisha capital if the Bill was tabled, which led to the final withdrawal of the bill by the government. The draft Chilika Regulation Bill of 2002 was subsequently redrafted by CDA and is yet to be passed by the state legislature.

The Chilika Bill had vast and potential impact on the lives of fishers; and it is therefore expected that the fishers had sufficient knowledge and information regarding this bill. During the study, it was attempted to ascertain whether the fishers were aware of the contents of the Bill, to which we got a mixed response. While 47 per cent respondents said that they were aware of the Bill and its salient features, 53 per cent were not aware. To support the research questions, those who had only heard about the bill and were not aware of the features were considered as "not aware of the bill". This reflects that the fishers were not briefed on the contents of the bill and that it was prepared without consultation. This also meant that with the passage of time, the younger fisher people were no longer concerned with the old bill; and simultaneously did not comprehend it in its true sense. It also strongly endorsed that consultation with the fishers should be considered as an essential prerequisite, for the revised formulation of the bill in the future.

During the initial discussion with the Fishery Department Officials at Cuttack, Odisha, they argued that the situation of Chilika could have improved, had the fishers allowed the Chilika Bill to pass in the Assembly. However, a local leader from village Mahisha pointed out that

> Several outsiders who are opting for fishing as an occupation have brought rivalry to the local fishers and this has brought apprehension among the local fishers about their fishing rights. So the indigenous fishers feel that outsiders should not be allowed for fishing in Chilika. The proposed Chilika Bill

would have led to more troubles. The Bill would not only have crippled thousands of traditional fishers who depend on fishing for survival, but the government-sponsored shrimp mafias would also have spread their gheries all over the lake, damaging the surroundings and jeopardizing the lake. There is a need to have strict laws promulgated by the High Court, Supreme Court as well as the Odisha State Assembly; so that the lake could be cleared of all shrimp gheries immediately and the shrimp mafias can be thrown out. In fact, the Chilika Bill should never be passed in its current form as it will endanger the lives and livelihood of millions of traditional fisher people of the region.

The findings so far highlights, that rights of access and use are of great concern to the fishers in Chilika. With the change in the fishers' perception, they now understand that common property is not free-for-all as open access resources and even a well organized common property regime also requires consensus amongst all before any action is taken. As we have discussed at length in the previous chapter, for fishers, caste, religion, and social norms created a sort of protection of livelihoods by barring outsiders from entering in to fishing. However, with the passage of time, the influence of traditional resource sharing system by demarcating the occupational structure has weakened. Thus, depending on the manner in which the fishing communities of Chilika perceive their own rights of their access to the lake resources, determines the ease with which outsiders gain access to the same resource.

During the study, we took the views of the fishers on whether outsiders getting into fishing activity in Chilika were a threat. Around 72 per cent respondents feel that outsiders are a threat to traditional fishing community and should be kept away. On the other hand, only 28 per cent do not think that outsiders are a threat to the traditional fisher people. The local political leader of Bhusandapur traced the problem way back to 1991, when the proposed Chilika Aquatic Farm (a joint shrimp farming project by OMCAD of the Odisha government and the Tatas) was being executed. Academicians and environmentalists supported the fishers' objections and the protest took the form of an environment protection movement successfully compelling the Tatas to withdraw from the project. Yet, the departure of the Tatas did not necessarily lead to the ousting of the shrimp culture outsiders. For the time being there is no method for limiting the number of fishers in Chilika and if outsiders enter the fishery, the problem of competition reappears. During our first visit to his house, Benu Behera's wife mentioned

> ..people fighting within themselves for land and fish is common here. But the Tata Company's entry into the lake led to a situation when we started fearing about our life. Gundas (muscle men) from outside started coming here and we could see a nexus between the police and gundas. We had sleepless nights in fear of our lives .... but fishers from every corner revolted back. My husband also went to Bhubaneswar to take part in the rally. Now we do not have Tata any more, but we still have the problem of chungudi chasha (shrimp cultivation). My husband says there is no harm in doing this, but why should we suffer because of outsiders? The Sarkar (government) should think about us and should compensate us.

Benu's wife whom we met again during our second visit during March 2012 was looking worried and distressed. Her younger son was not happy with the fact that both Benu and his wife were not able to support the household related work and wanted them to leave home. Neither the police nor the local community were of any support to them. Benu's wife told us the police usually do not get into family-related issues and ask the local panchayat members to resolve them. When we tried discussing the subject with Benu's son, he became furious and wanted us to leave their house immediately.

The long-standing resistance movement in Chilika therefore has basically emerged out of conflicts over access to natural resources, or occupational displacement. There is considerable competition between the various lobbies of fisher people and non-fisher people, authorized and unauthorized shrimp culturists, besides locals and outsiders over the aquatic resources of the lake. When such situations arise, either it leads to short-term demonstrations or long-term social movements. In case of Chilika, the tragedy is the inability of the government to recognize Chilika as a common property resource which is not owned; and therefore, needs careful planning, development and management of government policies structured for the commons. The Chilika Bill was in no way an outcome of meticulous planning, or with due diligence to the status of the commons. The perception about Chilika was perhaps to remain a property that was not owned; and conversely, the Chilika Bill was a measure to make it a state property with potential to increase state revenue without taking into consideration the local sentiments. Implementation of the Chilika Bill would not have resulted in improvement of resources or the local fishers' livelihood.

Presence, Role and Impact of Mafias and Police in Chilika

Free access to common property attracts many outsiders, often poor, in search of easy earnings and their efforts get affected when the resource patterns including the access rights are reduced. As pointed out before, open access also ensures entry of big players to the sector who start influencing the micro scenarios by monopolizing access to resources often with support of state and marginalized traditional parties. In case of Chilika, the role of "Mafias" has been highlighted in several platforms. For example, a local newspaper in 2006 (*The Dharitri*, 2006) reported that the mafias have badly affected the fishery sector in Chilika. They acquired 4,130 acres of land on which fishing rights had already been leased to fishing communities. They had no consideration for the environmental and natural balance of the lake, and their quest for rapid profits severely spoiled and polluted the lake.

It would perhaps not be prudent to define mafias from an academic perspective and should rather be left to perception of respondents. "Mafias" in Chilika are considered to be the outsiders, who have encroached the fishing grounds for shrimp farming. Many other outsiders who have also hired the local crews for fishing are also called mafias as they earn much larger shares of the profits. There are a few who define mafias as muscle men helping the Mafia dons or their higher bosses. Mafia dons in the

Chilika region include political representatives and their associates, bureaucrats, wealthy shrimp traders and their agents. Based on our data, the respondents were distributed into four categories, based on their views on the impact of the mafia's involvement in the Chilika region. Six per cent respondents said that the mafia's role had a positive impact on fishing, whereas 80 per cent considered mafias had a negative impact. Only 4 per cent consider mafias had a both positive and negative impact. Close to 10 per cent said that the mafia had no impact on fishing. Our enquiry of "impacts" here mostly focused on "influence on livelihood". A few who considered the positive impact of mafias perhaps are either those who have been benefited in the past or see a likely benefit in future. The 80 per cent apprehensive response however proves that impact of the so-called mafias or local goons still looms large and has a critical impact on the fishers' livelihood. Most Primary Fishermen Cooperative Societies (PFCS) have leased fisheries from FISHFED, and subsequently sublet them to mafias for their family maintenance. Thus, shrimp mafias are still motivating the fishers by extending doles and appropriating fishing area illegally for gheries. We therefore argue for a stronger enforcement of regulations targeting shrimp mafias to stop illegal fishing and encroachment. For sustainable management of fishery resources of the Chilika Lake, the Primary Fishermen Cooperative Societies (PFCS) also require strengthening. There is also requirement for proper demarcation of lease areas of all fishermen cooperatives along with an effective conflict resolution mechanism involving regular and periodic inter-village and inter-cooperative societies meetings.

Along with the role of mafias, is intertwined the important role of the police. Traditionally, law enforcement has been viewed as a crime suppressive task performed by policemen and the police stations close to Chilika keep a vigilant eye on the law and order situations. As we have discussed, traditional fishers living in the villages around the lake have been opposing the construction of illegal gheries by shrimp mafia which has led to a lot of bloodshed in the past. One of the most important events related to this conflict was, as mentioned earlier in this chapter, in May 1999.

Following a clash between fishers and non-fishers, three fishers were killed in police firing at Sorana, a village on the Chilika bank. While the police justified the firing, the fisher groups as well as the human rights activists referred to it as inhuman. The police in Chilika have long been criticized for their passive role in terms of managing clashes between the fishers and non-fishers. At the same time, the police have been able to maintain law and order in the troubled areas. However, we received a mixed response in this case. 47.11 per cent respondents informed us that police do not play a vital role in maintaining peace, which implies lack of confidence in the local policemen. 31.78 per cent, on the other hand, justified the role of police in maintaining peace and harmony in the local area. 21 per cent however felt that the police do not have any role in their area, even though they do have a general presence to maintain law and order in a normal situation.

The local police officers consider that the police have been largely able to safeguard the traditional fishers' interest by clearing the gheries and sending back the shrimp mafias. Yet, some of the local leaders and fishers gave us an alternate and contrary version. According to them, successive state administrations have always protected the shrimp farms from demolition despite the repeated court rulings by the Supreme Court of India as well as the Odisha High Court. The lobbies of the powerful shrimp farm owners apparently have always managed to influence the government duty bearers from attending to the grievances of the local fishers. The fishers have lost faith in the police and administration due to their apathetic attitude and inaction towards the illegal activities of the shrimp farms, which have flourished over time.

Chilika has experienced several demonstrations and small conflicts against those who tried to impinge on the rights of traditional fisher-folks and scores of lives have been lost in the process. The nature and kind of demonstrations have been different in the past and have also changed with the course of time. While most of the demonstrations were livelihood based, the specific objectives have varied, based on environmental, access rights or sometimes even based on regional causes. The priority of the study is not to focus on the type of demonstrations or their specific aspects. Instead, it chooses to highlight the fishers' perspective on demonstrations and their impact on livelihood. The demonstrations in the lake areas have mainly been fought for protecting the rights of traditional fishers and to safeguard the age-old fishing activities in the lake. While some of the demonstrations have been peaceful, others have turned into bloody battles. The study therefore attempted to ascertain, whether participating in demonstrations or social movements were common for fishers and the extent of their involvement in such movements. A substantive 80 per cent of respondents informed us that they had participated in the demonstrations, whereas only 20 per cent replied in the negative. "*Chilika Maati payin, aame jibana debu*" (We will sacrifice our lives for the sake of Mother Chilika) is the slogan, some of the fishers mentioned during the interaction. Benu Behera mentioned:

> We enjoyed our fishing rights over Chilika peacefully almost up to 1985. But, the good quality of shrimps with heavy export demand led to increase of shrimp mafias in the area. In nexus with the local political leaders, government officials and some traditional fishers, the non-fishers gradually started encroaching into the fishing areas and spread wide their zero nets and shrimp gheries which led to demonstrations. We are now reaching a point when we have to be ready to die at any point of time in order to protect our rights to life and livelihood. Our protest symbolizes the fact that even if the state can adopt a globalized pattern of business development, underestimating the traditional fishery rights and sustainable fishery management, the resource misuse can potentially bring the traditional fishers together.

Traditional fishers, unlike their modern counterparts, have a close relationship with the lake and Benu's statement clearly highlights the fact that there is an anguish and unhappiness among the fishers living around the lake. The traditional fishers are organized and have better social relations and they understand the cause and effect relation of the issues, which is largely affecting the fishers. While discussing with one of the police officers regarding Chilika, we were told that there are no more demonstrations or social movements in Chilika. Very rarely tense situations arise, and they are mostly due to the mobilization of the local leaders. However, fishers from Gangadharpursasan village unanimously emphasized that demonstrations and local movements have continued in Chilika and numerous factors have contributed to them. With the increasing dominance of the business mafia, powerful industrialists, elite shrimp merchants and politicians, law and order has been adversely affected. Therefore, demonstrations leading to violence often occur if a situation goes out of hand.

Sixteen per cent respondents informed us that they are being mobilized by a political group for demonstrations. The major chunk of respondents, i.e. 77 per cent responded that the fishers themselves led the demonstrations. They feel that the political parties have their own objectives and thus they do not allow them to intervene in their issues. Only 7 per cent respondents informed us that they have no idea about the leadership and they follow their families' decision for participating in the demonstrations. When discussing the political affiliations, awareness or interest, to our surprise, most of the fishers provided incoherent answers. The Sarpanch of Langaleswar blamed the Congress Raj (after independence) for having brought about the principle of lease of the brackish water area, especially for the culture of shrimps. He referred to the zamindars of the British rule from whom they used to obtain most of the fisheries in the lake by paying a royalty. According to him the then law and order system was good, and the government used to take care of the fishers. However, another elderly person blamed the current government for putting the Chilika Bill for its approval, as the Bill had many drawbacks and would have spoiled the life of fishers. However, it is quite evident that irrespective of divergent opinions, none of the political parties have done justice to the fishers. Thus, each government has tried to support the shrimp mafias and therefore the level of dissatisfaction is high with regard to any of the political parties. This however ensured that the mutual anguish against the political parties have further brought social collectivism among the fishers leading to strengthening their age-old argument for 'gheri free Chilika'. With this the role of state has become important in terms of safeguarding the rights of traditional fishers and providing them with sustained livelihood opportunities.

The Chilika Lake currently faces diverse problems combining both natural and anthropogenic pressures, including siltation, decrease in fish productivity, change in salinity, increase in fresh water weeds, change in species composition of avifauna, eutrophication, and excessive exploitation of bio resources and overall loss of biodiversity. These have already resulted in and have further potential for degradation of the lake eco-system. These issues are not independent in nature and impact the lives and livelihood of the fisher people in a major way. On being asked about the most important problem affecting the lake's fisheries, more than half of the respondents considered it to be the permission granted to outsiders to participate in lease bids while 19 per cent respondents thought of the increasing number of mechanized boats as the major issue due to their impact on environmental pollution and to traditional fishing. Eight per cent respondents thought that over fishing in the lake is an issue and should be taken care of, whereas 7 per cent think that environmental pollution is a major threat for the lake as well as the fishers. Only 1 per cent said there are other reasons which are more appealing than these issues for the Chilika including breaking of community norms, natural disasters, and illness during the rainy season, or even conflicts among the group members. Our findings further reflect that the fishers are aware of the issues and are looking for solutions. Fishing has been a tradition in the Chilika Lake region, and hence goes beyond merely being a source of livelihood or a coping strategy practised by most rural households in times of crisis, seasonal change, etc. Leasing lake water for shrimp aquaculture is an issue which the fishers cannot simply accept. This issue, if kept unresolved, may lead to intensified protests. During the interactions, a few fishers further expressed apprehension that they might eventually be squeezed out of the sector if the lease policy were to be further strengthened. This however points towards the importance of creating decision-making structures at the community level, thus providing autonomy to the community in deciding conflicts.

Decision-making at the fisher community level follows a multi-dimensional approach. While family-related decisions are mostly taken at the household level, any decisions pertaining to resource use with implications on others are jointly taken with the local administration. This includes local police or by the community leaders, or even by the NGOs or the local political leaders including the panchayat members. Irrespective of the participation of local administration in the decision-making processes, the call of fishers hold significance. While the decisions by the administration mainly follows the set of rules and regulations established by the law of the state, decisions by the community also looks at the welfare of the overall community. Decisions taken at the community level are the results of several factors, including the type of decision-making structure, the composition of leadership, and the basic characteristics of those who accrue benefits. The findings from our study highlight that 54 per cent of respondents claimed to follow the decisions of the community and not any other authority, whereas 28 per cent replied to listen to the police and administration. Five per cent listened to the political leaders and 3 per cent gave priority to the NGO opinions. Interestingly, 9 per cent followed others including their religious leaders. It is quite common that in Chilika there is an informal judicial system in place, which takes care of the fishers' grievances. Most of the fishers follow their community decisions and act accordingly. According to the respondents, political leaders hardly visit their places. Only the local leaders visit their villages and that too mainly during the elections. It may be interesting to note that the villagers do not link their sarpanch (local Panchayati Raj representatives) with any political party but consider them as one of them. Therefore the role of political leaders in decision-making at community level is low. Also the fishers have very limited faith in the police and administration and therefore do not take their views seriously. Further, the traditional fishers' cooperative plays a limited role in collective decision-making, as many traditional fishers believe that the fishing cooperatives were responsible for the annual lease rights for extensive aquaculture operations using the gheri system, and the illegal leasing of their fishing rights to outsiders for a fixed payment in return. Thus, the fishers in the current context are seen to be influenced by a limited range of actors and authorities and mostly adhere, strongly connect and associate to their collective entity that has evolved and continues to hold a place of primary prominence in their lives.

Our argument therefore focuses on the lease policy of the Government of Odisha for shrimp culture in Chilika introduced in 1991 as the main cause behind the deterioration of the lake's ecology. Further, the socio-economic condition of the traditional fishers, eventually resulted in large-scale demonstrations and loss of lives. A total ban on illegal and unauthorized shrimp culture may increase the level of income source for fishers of Chilika. It is all the more important not to think about only the economic considerations but also there is a need to harmonize the social as well as ecological imperatives.

Conclusion

The study more particularly recognizes the importance of a system, one in which the people respect and follow rules that they make for themselves. When they are given the right to take decisions for the resources on which they are dependent on, they try and ensure that the resources are properly used and managed. It also makes them secure about their rights, thus encouraging them to take time to ensure that their management practices are enforced. Thus it becomes important to ensure that these informal rights have legal standing and recognition. Fishing in the lake, follows an age-old system of dividing boundaries which is otherwise difficult to be visualized. The traditional access-regulating mechanisms in Chilika Lake are mechanisms that have been evolved by communities to manage their fishery resources sustainably that would last for a longer term. The core of this system lies in the acknowledgement and understanding of the fishers' symbiotic relationship with their lake that they are dependent on, simultaneously also realizing the vulnerability of the ecosystem which they call their 'home'. However, in the recent past the fishery development in Chilika has largely focused on products and not on people; resulting in a situation where traditional fishery-related activities are seen to be taken over by a new group. More specifically the views of the traditional communities have been ignored and the fishers who were already hard pressed on account of highhandedness of the middlemen have further become marginalized. The state government, on the other hand, has ignored the quandaries of fishers leading to discontent among them. Social unrest has been brewing in and around Chilika over a period especially with relation to policy pertaining to shrimp gheries and modern commercial ventures which have threatened the survival of fisher folk.

The central theme of our arguments is based on the facts that justify the state should avoid impinging on the rights of the traditional fishers and instead identify its role to supporting the fisher community of Chilika. In articulating a framework which strengthens their traditional system and legalizing it to grant them more autonomy, along with providing them with necessary information for community decision-making. The government should not move forward, donning the role of a command and control authority by taking up the property rights over the common property resource, i.e. the lake. It further justifies community-led collective action within a clearly laid out framework of minimum government rules, regulations and enforcements, that can lead to efficient use and management of common property resources like that of Chilika.

NOTES

1. A policy is more a statement of broad goals or objectives that is made by the government to achieve a certain outcome and may not involve specific legislation, whereas forming the act is a legislative process and once passed by the legislative bodies, become a law. Acts once made are binding on people or governments whereas policy is not binding on governments to achieve the targeted objectives.
2. Gheries are areas encroached for illegal prawn cultivation inside the lake. These enclosures are made by threading fine-mesh nets, hooked to bamboo and wooden supports. These are erected for the culture of Bagda Chingri (local name for tiger prawns).

REFERENCES

1. Demsetz, H. (1967), "Towards a Theory of Property Rights". *American Economic Review*, 52, No. 2, pp. 347-379.
2. Klitzner, M.D., Vegega, M., and Gruenewald, P. (1988), An Empirical Examination of the Assumptions Underlying Youth Drinking/Driving Prevention Programmes. *Evaluation and Programme Planning* 11(3): 219-235.
3. Berkes, F. (1986), Local-Level Management and the Commons Problem: A Comparative Study of Turkish Coastal Fisheries. *Marine Policy* 10: 215–29.
4. Dragun, A.K. (2001), Common Property Resources. In *Encyclopedia of Political Economy*, ed. P. O'Hara, 118-120. London and New York: Routledge.
5. Ostrom E. (1990), *Governing the Commons: The Evolution of Institutions for Collective Action*. Cambridge University Press.
6. Harvey, D. (2011), Roepke Lecture in Economic Geography—Crises, Geographic Disruptions and the Uneven Development of Political Responses. *Economic Geography*, 87: 1–22. doi:10.1111/j.1944-8287.2010.01105.x.
7. Hardin, G. (1968), The Tragedy of Commons. *Science*, 162.
8. Hanna, S. and Munasinghe, M. (1995), "An Introduction to Property Rights and the Environment". In: *Property Rights and the Environment: Social and Ecological Issues.*
9. Bollier, D. (2007), The Growth of the Commons Paradigm. In Hess, C. and Ostrom, E. (eds.), *Understanding Knowledge as a Commons: From Theory to Practice* (pp. 27-40). Cambridge, MA: MIT Press.
10. Feeny, D., Hanna, S. and McEvoy, A.F. (1996), Questioning the Assumptions of the 'Tragedy of the Commons' Model of Fisheries. *Land Economics*, 72, pp. 187–205.
11. Dove, M.R. (1993), "A Revisionist View of Tropical Deforestation and Development". *Environment Conservation*, 20, pp. 17-24.
12. Berkes, F. and Folke, C. (1998), "Linking Social and Ecological System for Resilience and Sustainability". In *Linking Social and Ecological System: Management Practices and Social Mechanisms for Building Resilience*, F. Berkes and C. Folke (eds.). Cambridge: Cambridge University Press.

13. Levi, M. (1988), *Of Rule and Revenue*. University of California Press. Retrieved from http://www.jstor.org/stable/10.1525/j.ctt1pngtk.
14. Peters, Pauline E. (1983), 'Cattlemen, Borehole Syndicates and Privatization in the Kgatleng District of Botswana; An Anthropological History of the Transformation of a Commons', PhD Dissertation in Anthropology, Boston University.
15. Dharitri (2006), 651 of the Chilika is Under the Control of Mafia. mimeo. *Dharitri*, Bhubaneswar, November 13.

5
Balancing Livelihoods and Sustainability

This chapter plots livelihood patterns which are linked to the socio-economic aspects of fishers and enquires into some of the external and internal variables that impact fishery livelihoods. The premise that the means of making a living or to secure the necessities of life, ultimately translates into a livelihood defining, impacting and sustaining the ecosystem, has been comprehensively analysed. More specifically, the current status of the fisher community with respect to the personal assets, credit status, and the existing external support system has been discussed in depth. Historically, ownership of economic resources and physical assets has symbolized authority and power. Possession of assets has traditionally been a mechanism of ensuring cover or a protection against combative economic shocks, including unpredictable ones. This is associated far more deeply with social and cultural practices in all rural areas, specifically indicating material well-being. Asset ownership is critical to both resilience and escaping poverty (Shepherd 2010). Fishermen with no assets, become more vulnerable to economic crisis when earning from fishing becomes difficult due to a decline in fish catch or bad weather conditions. Those households that are able to produce some surplus spend it quickly during lean periods or use it on maintenance of households or for major social occasions. When the surplus is spent, fishers are forced to diversify or sell their household assets. Thus basic household assets not only help in generating some regular income, but also help in commanding additional resources during the lean periods.

Defining Livelihood

The Oxford Dictionary rather appropriately defines livelihoods as 'a means of securing the necessities of life'. It is best understood as the competencies, resources, actions, and the ability to mitigate stresses and shocks, for generating means of living and sustainable livelihoods (Chambers and Conway 1992). Sustainability is accomplished when livelihoods 'can cope with and recover from stresses and shocks, maintain or enhance its capabilities and assets, both now and in the future, while not undermining the natural resource base' (also Carney, 1998). The concept of livelihood in a broader perspective, based on the social sciences, contains in its domain social and cultural capital, cultural knowledge, information, social networks, legal rights, and assets such as land (Blaikie et al., 2004). The livelihood eco-systems with its varied components, creates the physical, economic, social, and cultural space, wherein the families live (Hogger, 2006). In actual practice, livelihoods consist of the group of actions which are influenced by multifaceted elements, that contribute to generate income, or secure food supplies, for an individual or group of individuals for their living. In order to understand the livelihood status of a community, it is important to understand the choice and strategies the community members follow to ensure their livelihood. Any analysis of livelihoods that emphasizes only the major economic activities in which the members of the group are engaged for living, will likely ignore critical social and cultural relations that characterize the fundamental aspects of their livelihoods.

Livelihoods: Facts and Figures

The fishery sector contributes in generating livelihoods, as well as supplementing the diet of a large proportion of the population worldwide. Inland fisheries account for about 15 per cent of total global employment (FAO, 2000) and hence is considered to be of immense significance to the rural masses, especially the poor. While some fishers are engaged in fishing full-time others engage in it as a part-time business, taking advantage of seasonal availability. Fishery-related activities provide livelihoods to seven million people in India and directly contribute to the national income (Government of India, 2000). However, during the last several years, increasing levels of livelihood vulnerability among fishers have been highlighted in different forums, and mostly attributed to an overall decline in availability of fish in terms of quantity, quality and variety. With the increasing number of stakeholders involved in this sector, the catches are spread thin and consequently, employment prospects have become increasingly bleak. Apart from issues of availability there is also the critical question of accessibility which has bothered the international community for the past few years. Large-scale development, exemplified by dam construction, port building, and the emergence of power corporations is resulting in the steady displacement of indigenous or local fishing communities.

Considering the larger livelihood framework, substantive research has taken place in the past keeping in mind the socio-economic conditions of small-scale fishers. Results have clearly highlighted the fact that overall fishery-related livelihoods systems are multifaceted, dynamic, and that the livelihood systems of small-scale fishers keep them trapped in poverty (FAO 1995; Rubinoff 1999; Sekhar 2004). The deep-rooted and complicated livelihood systems also lead to arbitrary fishing activities without compensating for environmental degradation. In order to understand the livelihood of fishers of a region, it is important to know the degree and extent of their dependence on the source of fishing. The dynamic nature of the environment for fishing, implies that the ability of people to sustain their livelihood, depends on addressing a host of shocks and challenges. They include uncertain fish catches, lack of security, possibility of natural calamities (such as cyclones or floods), lack of capital, alternate employment opportunities, and provision of health and education facilities. Open access to any part of their common property including a water body, while creating livelihood opportunities for large numbers of the poor is not an unmitigated blessing. Especially since accessibility can be adversely impacted by the imposition of regulations or restrictions by the state. The common access areas/properties like Chilika, also attract the high-stake players, prospecting for big profits at low investment. Their activities too, overtake the traditional population, and drive them into further marginalization and deprivation.

The livelihood of fishers living in and around the lake is entirely sustained by the highly productive lagoon eco-system and its rich fishery resources. On the other hand, the higher salinity and low productivity of its land, poses a serious handicap to the non-fishers of this area. In the past, salt was manufactured on these lands, but in the early 20th century, salt fields were converted into agricultural lands, which were open to crop loss by brackish water inundation from the lake. Rabi crops are the main crops of the locality, though in the northern part of the lake particularly around the bank of the Daya river, some people cultivate vegetables. Salt production is an alternate livelihood in neighbouring areas of Chilika, particularly in the southern part of the lake; however, it is found that the lease holders of the saline land are mostly outsiders. The local people are therefore forced to sublease the salt fields, and work as tenant cultivators or as farm servants of the lease holders or businessmen. Needless to say, salt production is a seasonal occupation that people can take up only in the summer. Dry fish is prepared in most parts of the lake wherein both male and female members of the fisher community engage in the business and together contribute some earnings to manage their families. Some members of the community sell their products by moving from village to village even under adverse climatic conditions. With prospects for alternative occupation severely constrained, fishers suffer economic adversities, including unemployment and forced migration.

Livelihood Asset Structure

Contemporary theoretical and policy discourses around livelihood, reflect the lack of assets as both a symptom and cause of poverty (Hoddinot et al., 2000). The *India Human Development Report* (2010) highlights household assets as everything that was owned by the household and had monetary value. Moser (2007) deliberating on the 'asset-based approaches' to poverty, highlights: "assets are a stock of financial, human, natural or social resources that can be acquired, developed, improved and transferred across generations.

It generates flows or consumption, as well as additional stock. In the current poverty-related development debates, the concepts of assets or capital endowments include both tangible and intangible assets, with the capital assets of the poor commonly identified as natural, physical, social, financial and human capital." The overall asset based livelihood context for fishers thus, can best be understood by first analysing the larger diversity of livelihoods of fishers, subsequently examining diverse asset types that impact the fishers' livelihoods; and finally by analysing the broader socio-economic and political climate, that decides the access of the fisher community to assets. For the convenience of our study, we took two kinds of assets, i.e. land, and livestock.

Land

Ownership and land use patterns in the rural landscape are traditionally important criteria used to evaluate the economic condition and output of a household. The *India Human Development Report, 2010* estimated that land along with buildings constituted more than 85 per cent of the total value of assets in rural India. Interviews and discussions with local fisher people who also cultivated their lands, were not only pertinent to socio-economic land use data, but contributed equally to land use characterization from various technical perspectives.

The pattern of land ownership displayed major facets of homestead and agricultural lands. Around a third (28 per cent) of the households owned homestead land more than 10 dc[1] while 24.7 per cent held land between 0 to 2 dc. Of the 30 per cent who owned agricultural land, all had holdings of more than 10 dc. The existing land use pattern of the households, given the high incidence of small farms, indicates that a shift from agricultural cultivation to shrimp aquaculture may generate the desired increase in income. Technically, this comprises the relative minority that stands to benefit from the dredging of a new sea mouth into the lake in 2002, according to Dujovny's (2009) study. Dujovny makes a case for how increased benefits for the agriculturists on account of the dredging, have greatly marginalized that part of

Table 5.1: Land Use Pattern in the Study Area

Land Type	Nil		>0 dc. to <2 dc.		>2 dc. to <4 dc.		>4 dc. to <6 dc.		>6 dc. to <8 dc.		>8 dc. to <10 dc.		>10 dc	
	Nos.	%	Nos.	%	Nos.	%	Nos.	%	Nos.	%	Nos.	%	Nos.	%
Homestead land	37	8.2	111	24.7	50	11.1	44	9.8	53	11.8	28	6.2	127	28.2
Agricultural land	298	66.2	8	1.8	-	-	5	1.1	-	-	-	-	139	30.9
Land leased in	434	96.4	-	-	-	-	-	-	-	-	-	-	16	3.6
Land leased out	437	97.1	-	-	-	-	-	-	-	-	-	-	13	2.9
Land operated otherwise	445	98.9	-	-	-	-	-	-	-	-	-	-	5	1.1

the community which depends exclusively on fisheries for its livelihood. For example, dredging has greatly increased the tidal flows in several parts making it doubly difficult for the fisher folk to cast their nets in specific seasons. On the other hand, it has resulted in the conversion of agricultural lands into shrimp farms, which poses both opportunities and challenges. Aquaculture can create considerable employment for villages adjacent to Chilika and its promotion should be aimed at securing alternate livelihood avenues beyond fishing in the lake. The primary challenge however remains to integrate efforts towards a sustainable model of production that does not alter or impact the natural balance of the ecosystem.

Livestock

Globally millions of people earn their livelihood from livestock as it provides direct cash income. Fishers who adopt livestock attempt to use them as buffer for their extant or expected resource availability, against the unexpected fall in catch. Therefore, in a fisher's family, it is almost imperative to own livestock, with a hope of earning some income during the lean period. We focused on understanding the possession of animal assets by the fishermen as an alternative source of livelihood. Possession of livestock in case of fishers also needs to be measured in terms of interactions between their future attribution and their current utilities within the overall asset possessions and livelihood approaches. However, the difficulties encountered in counting, prevented a detailed reckoning of the animals. We found that only 9.3 per cent of the families have cows, whereas 4.4 per cent own buffaloes. Similarly, 4.4 per cent of households have pigs and only 0.2 per cent have goats. Twelve per cent families also have poultry including chickens and ducks. The relatively lower number of households possessing livestock as found in this study, also contradicts the general and globally understood system of the rural poor, using livestock as a coping mechanism against shocks. This should be explored further as it would be an important aspect to consider while designing alternative livelihoods for fishermen in Chilika. The possession of a lower number of livestock might also be interpreted as the intractable and entrenched nature of the fishermen, when it comes to livelihood options. Various assets, including consumable, productive, and convertible can fulfil different projected or unexpected needs that may arise in the lives of the fishers. The dynamic relation between the fishers' asset base with different functions and livelihood activities provide a subjective well-being context, which requires further exploration and analysis. However, the current asset base of the fishermen poses a picture of extreme vulnerability as well as a challenge regarding introduction of alternative livelihoods. There also seems to be an absence of diversification and adoption of coping mechanism among the fishermen. This raises very relevant and urgent questions of sustenance and adaptation of the fisher people in the event of an alteration of the fishing resources in the lake ecosystem and their traditional fishing livelihood.

Market Attributions

Changes in business patterns in the recent past have brought several changes in the catch sharing pattern in Chilika. Several fishermen have shifted to selling fish collectively and sharing the cash returns. Though independent traders are not able to outbid local businessmen; but local cooperatives ensure regular and uninterrupted business for the fishermen. In Chilika, the middleman-traders are relatively new players, engendered by the growing shrimp export markets. The data speaks for itself, as the total shrimp export from India went from 55,736 mt in 1987 to 134,815 mt in 2002, and in Chilika, the annual shrimp and shrimp production grew from 1144 mt in 1985 to 4927 mt in 2013 (Ganpathy and Viswakumar, 2001, CDA and PIB, 2017)[2].

The fish procurement and trade networks underwent a major transformation during the late 1980s and early 1990s, in terms of reorganization following the boom in shrimp export. Consequently, fishermen reinvented their roles as middlemen by procuring fish for non-endogenous traders and processors. Our study found that 63 per cent of fishermen sold their catch through some middlemen, whereas 29 per cent sold their fish directly in the market and 2.7 per cent sold it to the processing units. This reflects that fishermen usually sell their catch to a middleman (who doubles up as the moneylender) from whom they borrow money at predetermined rates of interest. The interest rate varies from region to region, market to market, and season to season. Therefore, efficient facilitation of timely credit by the government can ensure better market linkage and incomes. Besides, sustainable supply chain development and adding value through developing small processing units in this scenario may ensure that the fishers undertake product grading of their catch. Mr. Mohanty from CDA stated:

> The Government of Odisha is considering providing fishers more options regarding selling their fish, though we are in an early stage and a lot of discussion has to happen for it to be successful. The first consideration is making fish available to all. There is a huge demand for Chilika's fish in the capital (Bhubaneswar) and we have to ensure quality supply.

Okonta and Ekelemu (2005) had outlined the significance of processing and preservation methods to address deterioration of fresh fish immediately after the harvest, the absence of which would lead to economic losses. Dearth of adequate fish handling, processing techniques and storage facilities also lead to non-availability of fish supply to poor rural households depriving them of an essential diet requirement (Ayuba, Omeji, 2006). In Odisha, several rapid changes occurred with introduction of cold storage in the 1990s in the fisheries sector; with important centres like Pentakota for fish landing, setup with ice plants that were being accessed by fishermen by 1988. Also, as noted by Dahl and Forsgren (1988), preservation of fresh fish with salt was being carried on board the Beach Landing Craft (BLC) from Pentakota. The last decade also witnessed a steady increase in road transport by both insulated and non-insulated trucks.

While we tried to understand the existing storage facility, to ascertain whether the fishermen have enough options available to store their fish, the findings reflect that while 45 per cent fishermen store the fish in their boats, before selling it or handing over to customers; 37 per cent store them in a godown, which is safer. Sixteen per cent store them in their own house and only 2 per cent store them in other places. This proves that the maximum number of fishermen sell their fish directly in the market. This is also being done through the middlemen, who instantly take the fish from the fishermen. The accessibility to storage facilities, is also dictated by their distance from urban centres. Hence a number of villages located in the islands of Chilika, are handicapped by the lack of storage facilities. New technologies such as cold storage and better transport will help the fish producers to obtain higher returns for their catch. Price of the fish will increase as ice as preservation resources will add to the maintenance and overall cost.

Impact of Shrimp Export on Local Livelihood

Chilika Lake is a major source of fish, shrimp and crab in the state. A total of 217 species of fresh and brackish water fish, 24 species of shrimp and prawn, 9 families of crab and 136 species of molluscs were reported from the lake (Ayyappan and Jena, 2000). More than 92 per cent of the shrimp produced in the state of Odisha are exported abroad or to other states as part of a trade liberalization agenda that has unfolded in India since the 1990s (DOF, 1998: 14). The major international markets being Japan, Middle East, Southeast Asia, the United States and Western Europe (Marine Products Export Development Authority (MPEDA), 1995: 2a; DOF, 1998). During the last few decades, shrimp aquaculture has overtaken the traditional systems of fishing and the lake has become one vast shrimp farm, attaining the dubious distinction of being dubbed, "the major illegal aquaculture complex". The spurt in interest over shrimp culture has been fuelled by the twin factors of low investment requirement and a huge global market; which in turn aided by non-requirement of traditional knowledge, has allowed local politicians to lease vast areas of the lake. It has also led to bloody battles in the lake area as the gheries used for shrimp culture block the water channels, affect the migration of fish juveniles, cause loss of grazing ground for fish, and adversely impact the indigenous fishermen of the area. In 1997, the Odisha

State Assembly House Committee agreed upon and affirmed the adverse impact of shrimp aquaculture on the living conditions of fishermen as well as large encroachment into lake areas resulting in exploitation. However, so far no concrete step has been taken in addressing these matters in Chilika.

The impact of shrimp aquaculture is not a simple binary issue; its complexity requires careful consideration before being taken forward. Our data and interviews revealed that nearly 43 per cent respondents felt that shrimp export helped local livelihoods, against 47 per cent respondents who did not feel the same, while 10 per cent respondents had no idea about the debate. Thus, the numbers for and against the promotion of aquaculture, are more or less equally matched. Interestingly, the study establishes a direct link between the arrival of big exporters and growth in the export market. It creates a new hierarchy among the traders, based on level of investments and commodity. Thus, traders with deep pockets dealing in export varieties of fish are placed high up on the ladder, while petty fish traders occupy the lower rungs. This stratification in turn gives rise to two points of view; that the high end exporters do not add much value to the local economy, and conversely that even a small number of exporters, help in strengthening the local economy. The latter is engendered by exporters engaging local fishermen in packing, loading and procuring packing materials like ice and thermocol baskets from the local market; thus, contributing to the consumer base for the local market.

The immediate attraction of a cohort of people with the perspective of gains for their community due to export led employment generation is unfortunately myopic and does not take into account the long-term repercussions and changes in the overall livelihood prospects. Though aquaculture has the potential to enhance profitability and productivity, but it does not necessarily result in greater job generation for the larger rural masses, unlike traditional fishing. This is indubitably stated by Sebastian Matthew, who heads the International Collective in Support of Fish workers (ICSF), "for every five or ten jobs created by capture fishing, you will create one job in aquaculture."[3] Thus, an increase in the number of exporters and rampant promotion of aquaculture is eventually going to result in high rates of unemployment in the long run. Furthermore, narratives from our field research show how the job characteristics change dramatically with the arrival of the exporters. Fishing communities either become feudalized or reduced to wage labour with the arrival of big businesses. This greatly corrodes all that was held to be of significance in cultural terms for a community. Such transitions will impact the psyche of a community whose rich history is reduced to 'wage hunting' (Breman, 1994), and as auxiliary assistants of export enterprises.

Fishing Merchants

Owners of mechanized and motorized boats are usually considered a higher class among the fishermen. Procuring a large boat and owning its fishing operation requires enormous investment. The mechanized boat owners are the local younger fishermen and powerful elite, besides new entrants from outside, with their own crew, who fish for them. The big traders hire locals by paying standard wages during the culture season, in addition to awarding extra remuneration following a surplus. However, these wage earners represent the people who engage them and are no longer representative of self-employed fishermen. This needs serious consideration as there could be large-scale exploitation in the fishery sector under the pretext of providing employment. This exploitation can take the form of disproportionate working hours, lower pay, or even abuse at the workplace. However, our interpretation of the basic facts is based on the fishers' perception and not from the other side of the picture. Sixty-seven per cent of the respondents informed us that the aquaculture farms never care about the local people and keep their own people; whereas about 31 per cent said that the farms prefer to employ local people. This data endorses the increased vulnerability of the local fishermen when confronted with limited employment opportunities. Our study also reflected 76 per cent of respondents feel the need to establish traditional fishermen's association, to further strengthen their voice for their demands. In view of the constant threat from the outsiders and big businessmen, the need of associations to safeguard the traditional fishermen becomes much stronger. Pallabh Maharana from Puri told us:

> There is an association comprising of only the traditional fisher communities as its members who work for the rights of fishers called "Matsyajibi Mahasangha". However, it must be further activated, strengthened, and to be linked with the government agencies like Chilika Development Authority. The agency should have the power to plan and implement conservation measures collectively. A sense of belongingness will bring more support towards both conserving and developing the fishery resources in the lake.

The Indian federal governance system has a variety of central and state level welfare schemes, with the potential to increase the fishermen's ability to adapt to change. The skill based training programmes can help in acquiring transferable skills which may result in more work opportunities. In the fishing sector, the trends currently point towards a gradual erosion of regulatory mechanisms that controlled common property lands in the past. An additive factor is the stress created by population increase on the extant resources and lake ecosystem. Access rights and control over common property resources is always an outcome of an indigenously evolved management system by local communities. Thus historically, access rights over Chilika have been managed by the fishermen; but over time, several government departments have intervened. It is interesting to note that out of the total respondents, 66.6 per cent wanted the community to continue regulating their access to Chilika, whereas 26.6 per cent wanted the state government to regulate the access. The rest responded that the lake should be regulated by both government and community.

Evidently a powerful indigenous management system is in operation in Chilika. This has supported the fishermen for generations and has proved to be beneficial for them. During informal discussions with some of the educated respondents from Mahisha, we were told that restriction of access to resources in the designated areas have affected fishermen during the last few years. A high degree of ambivalence is apparent in many of the purported restrictions. It is not uncommon for boats to be seized along the coast of Odisha, particularly the Gahrimata beach (*ToI*, 2015), on the pretext of protecting the habitat and nesting grounds of the Olive Ridley sea turtles; while concurrently, the state and central government have sanctioned construction of a port that is bound to have disastrous ecological consequences. Analogous situations have discouraged fishers from government regulation and encouraged support to open access. A common resource like Chilika Lake that is vital to the fishermen, should be central to development strategies that foster community ownership and management. Consequently, the authorities involved, should identify and highlight aspects like mechanisms of equity issues, understanding of evolution of community management over time, and the key elements of traditional management structure for a holistic outcome.

Coastal Odisha is rich for its fishery resources; however, it is not free from the problems of environmental degradation, over fishing, traditional fishing patterns, and conflicts between the fishermen regarding ownership. The replacement of traditional use patterns formal systems by regulation induced ownership is aggravated by internal disagreements within the latter; which causes increased conflicts. Thus, the very first of the eight principles articulated by Ostrom (1990), namely 'clear group boundaries,' for the successful management of common property resources, is the biggest hurdle, and maybe impossible in the context of habitats like Chilika. It is generally perceived that natural ecosystems are destroyed when the state regulates, restricts or closes off access to everyone, including the traditional users; claiming extensive ownership over the natural resources. Conflicts arising due to indistinct roles of multiple players adversely affect the access of people, who are dependent for their livelihood on the traditional resources. In the recent years, the access rule in Chilika has become more open as several new stakeholders have started claiming ownership and access on the lake waters. These include various state government departments, several NGOs, and political parties that impact the life of the fishermen. The study, based on the input from the fisher community revealed interesting perceptions about which department of the government controlled Chilika. Thus 41 per cent named revenue and the local administration, 16 per cent indicated forests and wildlife, 15 per cent said fisheries, while 19 per cent were clueless and 9 per cent cited other departments including the police. The Department of Fisheries is interested in higher revenues by way of granting more leases to outsiders for aquaculture, whereas the Department of Forests and Wildlife is interested in closing some areas for better control. It is thus evident that at a broader level there is a conflict between different state agencies due to opposing interests. It would be important to highlight a positive development cited by the respondents towards conflict control; following several bloody fights and subsequent interventions of the Odisha High Court, the Police have become active in the neighbouring areas of Chilika. Though the CDA has taken up some alternate livelihood generation activities for the fishermen but they have not yielded any result. In all this, the need towards resolution of issues related to access rights and control over the lake resources are not explicitly seen by the state as issues to be addressed immediately.

The role of the Chilika Development Authority for development of fishermen has always been under review by the researchers. Consequently, the responsibilities of key stakeholders including Odisha State Fisheries Cooperative Marketing Federation, Odisha Fish Seed Development Corporation, Odisha Maritime and Chilika Area Development Corporation Ltd., Central Fishermen Marketing Cooperative Society (Chilika), Directorate of Fisheries and various fish farmers' development agencies in the fisheries sub-sector have been noted by Fishers and Mahajan (1997:135). The CDA has, since its inception, prioritized the protection of the lake ecosystem by preserving its genetic diversity through various developmental activities, either implemented directly or through other agencies. It has also collaborated extensively to conduct surveys, develop plans and proposals for Integrated Resource Management in addition to working substantially towards control of silt load in the lagoon through catchment management, sustainable fishery resource development, conservation of biodiversity, eco-restoration of islands including habitat improvement of birds, promotion of eco-tourism, etc. But the science and logic that the CDA operates with may not necessarily be open to community participation whereby local knowledge is rarely factored into the modicum of research adopted by the body, and there could be imminent dangers from the exclusion. This is notwithstanding the fact that the CDA has taken up many developmental activities in the fishing villages adjacent to the lake, which are centred on enhancing the economic condition of the people. Besides, for its achievements in environmental protection, it had been honoured with the Ramsar Wetland Conservation Award and Indira Gandhi Paryavaran Puraskar in 2002.

During our study, we tried to understand, the people's perception of the Chilika Development Authority for the development of fishermen. This helped us comprehend the linkage between the traditional fishermen, the CDA and the kind of benefits the fishermen clearly expect from CDA. While 45 per cent respondents felt that CDA plays an active role in development of fishery and fishing activity in the lake, almost 55 per cent respondents felt that the CDA does not fulfil most of its designated objectives. CDA does not consider the control of access to the lake resources as an important factor that needs to be addressed on priority, which has created a clear level of discontent among the fishermen. While CDA has done more work in the environment sector, fishery development and community interaction have not garnered the requisite attention. The CDA currently lacks the required dedicated endeavour to empower the poor fishermen by strengthening their voice and fostering democratic accountability. It is important that the state should focus more on the socio-economic issues and factor in environmental issues, instead of doing the opposite. The CDA, in close collaboration with several national and international agencies including the Japan International Cooperation Agency, has brought about a degree of improvement in the community at the periphery, through the initiation of alternative livelihood activities. Now it will be vital for the agency to see how the best practices which have emerged can be replicated in other villages, so that the fishermen get an opportunity to garner more income.

The Government of Odisha has introduced a number of schemes for both culture and capture fishers in the state. Similarly, several schemes have also been introduced by the Department of Animal Husbandry and Dairy, Ministry of Agriculture, Government of India. While some of the schemes are applicable for the fresh water fishery, others are applicable for the marine fishery and brackish water fishery. Though there are no special schemes for the fishermen of Chilika, the common schemes applicable for the other fishermen also apply to them. Our study revealed that 95 per cent fishermen are not aware of the major government schemes for the fishermen and only about 5 per cent know about the schemes. Lack of awareness among fishermen about government schemes can be apprehended as deprivation of often vital, entitled and adequate facilities. Over the last five decades, compared to the traditional approach of fishing, which was largely labour intensive, great technical diversity has developed with modern, capital intensive, and specialized technology-based approach; and therefore, technical support for the fishing activity can always be an added advantage for the fishers.

A closer look at the micro environment reflects that fishermen are still using traditional methods of fishing or sometimes gear which adversely impact the population of fish and other species. There was also a concern that increased pressure on the same area has affected both the quality and availability of fishes. The study reflects a rather poor status in terms of technical support for the fishing activity by the government. Ninety-eight per cent respondents feel that technical support by the government is zero. Thus, the fishermen remain ignorant and bereft of the technical expertise required to improve their livelihood, or increase their productivity; as in value addition for fish catches or use of fishing new gears, motorized boats, etc.

With its inception, CDA has undertaken a few capacity building initiatives for the fishermen of Chilika. In collaboration with other agencies like NETFISH (MPEDA), CDA initiated the capacity building training for the local fishers with effect from January 2010, targeted to reach the entire Primary Fishermen Cooperative

Societies. During the financial year 2010, around a hundred training camps in 86 fishing villages covered 31 Primary Fishermen Cooperative Societies. Fishing is a traditional business and almost all the fishermen learn it either from their family members or their peers; thus very few of them get further training. However, most of the training is meant for fine tuning the skills they already possess. Sometimes the training is provided on the new technologies including driving motorized boats, using new sets of nets and fish breeding. Our study reflects that while 11 per cent fishermen are trained on some techniques related to fishing, 89 per cent fishermen are not trained on any aspect. With implementation of modern technology in the fishery sector, fishermen without proper training may find themselves obsolete in future. This will not only decrease their productivity but will adversely impact their livelihood. The most important training programme, for the fishermen are post-harvest handling and preservation techniques. This training is not only important for the fishermen but also for the fishing women, as they play an equally important role in fish handling. Similarly, local fishing communities should be trained on how to observe and conserve their coastal resources and biological diversity, besides season specific requirement and use of fishing resources.

Credit System

The credit systems in general follow a two-dimensional approach: formal and informal with the latter playing a predominant role in rural areas. The informal finance was often traditionally viewed as anti-developmental, exploitative, consumption driven rather than investment based, and incapable of expanding its volume and range of financial services (Pischke et al. 1983). Conversely, the UNDP (1997) highlighted the significance of informal credit institutions at the grass root level and its simplified access to credit through ease of applications, swiftness in extending credit, and focus on the local market. It also provided larger loans based on successful repayments, thereby addressing the need of the poor clients by attributing more importance to reputation and goodwill than collateral. Lamberte (1987) classified the informal credit system into five sub-sectors: Rosca (rotating savings and credit associations), landlord moneylenders, traded miller moneylenders, farmer lenders and professional moneylenders. Kwack, et al. (1981) classified the rural credit markets into friends, relatives and neighbours, farmers, landlords, input dealers, output dealers, professional moneylenders, pawn shops and Rosca. Credit essentially does not always involve cash; rather the transaction can involve barter economies as well, based on the direct exchange of goods and services (Ingham, 2004).

The modern banking system in India started in the colonial period, when the moneylenders were the main source of rural credit, which in turn limited the spread of the banking system. Rural credit disbursement of the moneylenders was characterized by the high rate of interest with high exploitation of the poor. In the early 1950s it was found that the moneylenders, traders and rich landlords accounted for more than 75 per cent of rural credit source (Shah et al., 2007). The immediate action from the government was to concentrate on Cooperative Credit Societies and by 1971, the share of cooperatives in rural credit rose to 20 per cent (Shah et al., 2007); though the latter faltered on their many shortcomings. The mutual support between savings and credit functions, vital for successful functioning of the cooperative societies was lacking in India. Considering these objectives, the banks were nationalized in 1969, subsequently, the provision of credit to the peripheral areas and needy was brought under a formal government policy (Ramachandran and Swaminathan, 2004). In the early 1980s employment generation and poverty alleviation were recognized as important objectives in the government policy. The society credit and the Integrated Rural Development Programme were created for initiating productive income-bearing assets among the poor through allocation of subsidized credit. But overall the phase between 1969 and 1990 saw the unprecedented growth of commercial banking in terms of geographical and functional reach (Shetty, 1997 cited in Ramachandran, V.K. and M. Swaminathan, 2004). The period between 1990 and 2005 saw mergers and swapping of rural branches, with the share of agriculture in total bank credit dropping from

19 per cent to 11 per cent (Shah et al., 2007). In a separate article I have (Ray, 2013), attempted to highlight the usurious credit arrangements, controlled by local moneylenders and commission agents, which continued to be rampant and fully operative despite formal credit institutions. The article argues that "the fishermen, who take loans primarily for economically productive purposes, effectively yoke their earnings from the next season's fishing. This finally limits their marketing opportunities, to place the catches at a remunerative market of their choice." One of the main reasons for continuing with this system of bondage, is the red tape involved in the formal credit sector. Thus the "opportunity costs" for negotiating with such institutions exceeds the interest rates charged by local moneylenders.

Incomes of the fishermen being seasonal, credit availability is required to smoothen out the irregularity between the flow of earnings and level of expenditure. Even seventy years after independence, several rural areas in India have no guarantee of state provided education and health. The public distribution and social security systems are also weak and are characterized by inefficiency and corruption. Social obligations also cast a heavy load on the fishermen. Each of the basic needs of health, education, food and social security, apart from the working capital and long-term investment requirements of rural livelihoods, create a major demand for credit. One of the factors accentuating the dependency on others is that there is often intense competition amongst fishers in Chilika. The challenge is to transform such competition into cooperation through socially viable institutions.

During the research an attempt was made to find out the opinion of fishermen on the availability of financial support, when required or in an emergency. If adequate financial support is available from a reliable source, it can lead to sustainable livelihood structure in any particular society. A maximum number of fishermen related the ease of procurement of loans or financial support from moneylenders; thus, an astounding 78.5 per cent of the respondents think that taking money from local Mahajan or the moneylenders is most easy. About 5.3 per cent feel that getting support from the bank is easier, whereas 6 per cent feel that getting money from family members is most easy. Only 4 per cent feel that getting financial support from the government is easier and 2 per cent feel that getting it from co-operatives is easy. About 3.6 per cent however feel they can get monetary support from other sources. The study reflects that, though the fishermen have enough sources of getting financial support, however the easiest source is to get it from the moneylenders. This re-establishes the notion that fishermen are prone to seasonal poverty, which leads to taking loans from various traditional moneylenders, who either charge very high interest or take the fish catch against the loan. According to Samal and Meher (2003), the rate of interest in case of the informal sector is about five times higher than that of the formal sector. On account of the failure of formal credit institutions to supply credit to meet requirement of households engaged in fishing activities, they are forced to borrow money at a higher rate of interest from the moneylenders despite a satisfactory per capita income. This deal comes with a condition to sell the catch at a predetermined price which is exploitative. Thus, a boat owner indebted to a trader is vulnerable, as he is often forced to sell his catch at the rate demanded by the trader.

An important indicator of poverty is the source of credit. Fishers are often devoid of assets or income for collateral which diminishes their credit worthiness from the perspective of formal lenders including banks and professional moneylenders. As a result they have to source credit in the form of loans from friends, relatives, local moneylenders, etc. Having an individual bank account is a sure sign of affluence, although not having an account is not necessarily a sign of poverty. Schrader (1992) explained that the literature of sociology of development labelled moneylenders with two stereotypes for the most part with one painting them as 'loan sharks' with the negative archetypal figure of oppression and exploitation over the poor peasants. The second stereotype considers them as traditional institutions which prohibited development and should be abolished to be replaced by banks and other formal savings and credit institutions.

Historically moneylenders in India have played a critical role in both capital and credit markets by aligning with the traditional and modern financial system and are variously known as Shroffs, Seths, Sahukars, Mahajans, and Chettiars in different parts of the country (Das-Gupta, 1990). By using working capital of their own, they primarily deal with providing small-scale financial loans with a substantial interest rate, and unlike formal financial sectors, do not get deposits nor ask for savings from others. While in most cases they provide loans against securities such as gold, jewellery, land, promissory notes, etc; in some cases loans are granted on personal endorsement and assurance of persons well-known to them. With low paying business and minimum asset base, fishers are one of the most exposed targets and frequently enter into manipulative dealings with moneylenders, averting them from capitalizing and scaling up their current occupations. In the fishery sector, intermediaries per se play pivotal roles by providing timely loans to the fishers for procuring or repairing of boats, nets, etc. and subsequently buying the catch from them at a lower price. While this works as a two-way principal and interest settlement system, it simultaneously creates an exploitative society by leveraging an unfair market structure. Most of the time, fishers also take loans from the same intermediary during a crisis or to cover the expenses of social occasions, resulting in a vicious debt circle. In other cases, fishers also borrow boats or nets from rich individuals, who are known to them and end up paying a sizeable amount over a long duration. The high interest rate and pledging some assets as securities also make the fishers susceptible to risk. The moneylenders usually follow the safe transaction model by not providing any physical documents to the lenders, making the whole transaction dubious and uncertain. This leads to a situation, where moneylenders can easily trap the fishers. Our study observed that around 50 per cent of the respondents have taken loans from the moneylenders, whereas the others have so far not taken any loan from the local moneylenders. The rates are dependent on the amount borrowed and the type of borrower. Kanda Allaya of Bhusandpur village of Khurda narrated that

Fishers seek financial support during a crisis and it depends on the seasonality and the catch. The financial support from the banks comes with low interest rates, but the informal moneylenders usually charge more interest. Most of the loans taken by the fishermen are from private moneylenders and not from nationalized banks. Often the fishermen are forced to deliver the catch to the moneylenders in exchange of the interest. Therefore, the traditional fishermen from Chilika have little say on the price of their catch in the market. They have become hostages in the hands of moneylenders who advance funds to them for their boats, nets and other needs and these moneylenders now control the fishing market. In fact, the tedious paper work and need of a land patta makes most fishermen unsuitable for loans from nationalized banks. Efforts should be made to ensure better financial services for the fishermen who are in need and it can be provided by establishing grameen (village) banks and cooperatives. This will not only ensure a healthy financial structure within the fishing community, but will also ensure that the fishermen have alternate livelihood options through linkage to effective credit structure in the area.

During the study we observed that the informal credit structure with high interest rate from the moneylenders to be one of the biggest bottlenecks in shaping the fishing community's long-term livelihood. However, while discussing this issue with some of the older fishermen; we learnt that despite the high interest rates the fishermen preferred borrowing money from moneylenders rather than formal institutions. This was because the cash provided by moneylenders is available when it is most needed. The moneylenders also do not maintain regular business hours, and usually work throughout the day, making themselves available to the local fishermen at any time. At the same time the moneylenders are easily accessible, being from the same locality and they maintain a close relationship with the fishermen which is important in case of financial transactions. Our arguments towards providing concessional loans to fishers is based on the notion that farm-based savings activity in the short run does not always offer a return which allows regular servicing of loans and at the same time meets the least possible consumption requirements. Loans from formal financial systems with low interest rate can help fishers in acquiring their own equipment including fishing gears, nets, etc. as well as help them in addressing the risks during distress periods.

People who belong to the lowest income group spend most of their income on food. It is true that mere extension of financial services to the poor fishermen will not reduce poverty. The purpose of the loan (i.e. for the purpose the loan is used), the literacy level of the head, availability of non-farm employment opportunities in the surrounding area also play an important role in it. Fishing households with improved access to credits will be in a better position to adopt technology, increase their incomes, and improve quality of food intake than those who do not have access to credit. Most of the time fishermen buy either the boats or net for fishing. In addition, they also buy other things like ice and baskets for the purpose of storing fish. It is therefore necessary to establish a system of procurement of fishing equipment on credit with very flexible conditions, without which the fishermen will inevitably fall into the trap of moneylenders. Our study reflects that 71 per cent fishermen do not buy any equipment (fishing) on credit, whereas only 30 per cent fishermen buy the fishing equipment on credit. This proves that probably because of the high interest rate or lack of credit structure the fishermen do not buy the equipment on credit. The other reason, which can lead to this situation, may be either a small price of the equipment, for which the fishermen do not feel like taking loans. Conversely, they may not risk taking high value loans to procure high priced equipment. Also, as a maximum number of fishermen work as hired crews, their basic equipment like boats and nets are provided by the boat owners, and they have no requirement for personal procurement. However, this clearly reflects a market without proper credit linkage and subsidies for the fishermen.

Moneylenders, who advance funds for mechanized boats usually lay claim to almost all the fish catch from these vessels. This leads to a situation wherein trade in fish is being increasingly controlled by a handful of powerful individuals, or their agents. In this situation they can literally dictate the price of fish, by controlling the auction mechanism or by demanding the entire catch for the funds already advanced. The middlemen trading in fish products also exploit the gullible fishermen by making payments once in a fortnight or month. But they get payments from their paymasters, higher-up in the trading network at least twice a week. "In a way they reinvest the money meant for payment of fishermen as loans," alleges Mr. Panda from Mahisha village. This can only be checked, if proper mechanisms of providing loans are available. Mr Panda added:

> There are two types of loans available in the areas; one-day loans are provided for a day and returned the same day or the day after, and interest was charged accordingly and the second one is a loan for a particular period. There are various sizes of loans (ranges from Rs. 100 to Rs. 50,000), duration of credit and the interest rates which are dependent on the amount of credit and the type of borrowers. The banks usually provide start-up capital to initiate fishing activity, whereas the moneylenders provide working capital to meet recurring costs in the activities as well as loans for consumption purpose.

In Chilika traders advancing money to fishers in return for fish is an old tradition. However, most of the time fishermen also give credit to the customers against selling their catch. In this case credit is linked to the buyer's ability to repay, or rather, to the fishermen's perception about the buyer's ability to repay. While giving credit, it is sometimes anticipated that the credited amount may not be repaid and thus could become a bad debt. In case of fishermen, if he gives a vast amount of credit, it may so happen that, he has to take a loan for his own survival. Our study reflects that 52 per cent respondents do not give credit to the customers, endorsing that there has been an important change in the fishing sector regarding the increasing monetization of transactions. With the arrival of outsiders, fishers cannot afford to, or do not need to, sell their fish on credit. It also re-establishes the fact that as the fishermen get loans from the moneylenders, they sell their catch to them only, and are left with limited scope to deal in the open market.

Social relations in a fishing village traditionally influence economic activities. For instance, fishers would generally sell their fish to women processors and traders. On special occasions, fishers would give time to the processors to dry fish and take it to the market for sale before expecting to be paid. In effect, then, processors did not need to invest anything up front. Similarly, the processors give the dry fish or the processed fish to shopkeepers without any expectations and get the money after the sale of the fish. Without the formal credit linkage, this whole social structure is open to high levels of distress. The traditional credit structure at the fishing village symbolizes a value chain, which usually results in setting up a strong bond between the communities. Proper channel of credit among the community members can ensure livelihood for many, and most importantly the women and most disadvantaged in the area, who usually do not have any capital.

During economic transactions, interest rates are decided between the giver and taker, or in case of a formal system, interest rates are pre-decided. This is not applicable in case of exchange of food items. The whole supply system in the rural areas is systematic. While most of the small grocery shops and vendors are from the same locality, items are provided in credit with a proper account keeping system. The rural families usually pay the amount on a periodic basis, mostly weekly or monthly. The shopkeepers usually charge a small amount on credit, which is negligible. This system of transaction is common in the villages of Odisha. As already discussed, social relations in fishing villages traditionally influences the economic activities. Our study therefore attempted to ascertain whether the fishermen charge extra for giving credit. While it was revealed that the fishermen give a small interest on the money that they borrow from the middlemen, it is indeed important to understand, when fishermen take an extra amount, they also provide the catch as credit. The findings from our study reflect that 87 per cent of the respondents do not charge extra for giving credit, whereas only 13 per cent charge extra for giving credit and this extra could be another 5 to 10 per cent of the settled amount. This proves that credit in the fish business is not linked to interest and is more informal. Fish has a diminishing value, if not sold fresh, thus fishermen are bound to sell them on credit. Similarly, for the fish workers involved in production and trading operations that require investments, the returns from one cycle of operations finance the next. This means the fishermen often have no option but to sell their products at whatever price is offered, rather than waiting for a better opportunity. The study reflects that while maximum numbers of fishermen actually sell their fish to the middlemen, the rest, who are independent, have to sell it in the open market with credit and without interest.

Conclusion

This part of our study highlights how the fishermen have been subjugated by the external forces–middlemen and moneylenders. After looking at the overall livelihood scenarios in the lake including land, livestock and the market structure, our study highlights the fact that both the social and economic structure in the lake is going through a transition, partially due to the entry of new generation fishers who are more informed and look for faster growth. Our field interactions reveal a distinct informal financial market, where entities with commercial interest provide credit facilities. The fishermen, who take loans primarily for economically productive purposes, are effectively kept in permanent bondage by pledging their earnings from the next season's fishing. This comprehensively limits their marketing opportunities, to place the catches at a remunerative market of their choice. The outcome is unsustainable use of open sources under regularly constrained situations. The cardinal issue is the coexistence of a distinct informal financial market in the presence of a distinct formal credit market. It is fascinating how the moneylenders make their presence felt amidst the formal and structured institutions. Especially when the latter operate on the business strategy to lend, a pre-determined budget, investment planning and with a responsibility to be responsive to clients' needs (social responsibility). Without a strong socio-economic foundation, the debtors may find it difficult to operate the fisheries that are highly exposed to constant changes in the environmental and socio-economic sphere. Putting it all together, we argue to have a more regulated credit structure in the region to ensure a decent livelihood for the fishermen.

NOTES

1. 1 decimal is equivalent to around 450 square feet. The larger household areas we witnessed during our survey was mostly occupied free lands.
2. Though there is a lack of Chilika specific data,"India exported 11,34,948 MT of seafood worth an all time high of US$ 5.78 billion in 2016-17 as against 9,45,892 tons a year earlier, with USA and South East Asia continuing to be the major importers. Frozen shrimp maintained its position as the top item of export, accounting for 38.28 per cent in quantity and 64.50 per cent of the total earnings in dollar terms. Shrimp exports increased by 16.21 per cent in terms of quantity and 20.33 per cent in dollar terms. Frozen fish was the second largest export item, accounting for a share of 26.15 per cent in quantity and 11.64 per cent in dollar earnings, registering a growth of 26.92 per cent in terms of value (PIB, 2017)." Source: PIB: Press Information Bureau Government of India, Ministry of Commerce & Industry, June 7, 2017.
3. TDC: Small-scale fisheries: interview with Sebastian Matthew, Executive Secretary of International Collective Support of Fisherworks, https://www.befair.be/en/content/small-scale-fisheries-interview-sebastian-matthew-executive-secretary-international, accessed on July 14, 2017.

REFERENCES

1. Chambers, R. and Conway, G.R. (1992), IDS Discussion Paper IDS Discussion Paper. Brighton: Institute of Development Studies, University of Sussex.
2. Carney, D. ed. (1998), Sustainable Livelihoods: What Contribution Can We Make? London: Department for International Development (DFID).

3. Blaikie, P., Cannon, T., Davis, I. and Wisner, B. (2004), *At Risk: Natural Hazards, People's Vulnerability, and Disasters*. New York, NY: Routledge.
4. Hogger, Ruedi (2006), Understanding Livelihood Systems as Complex Wholes. In *In Search of Sustainable Livelihoods: Managing Resources and Change*, Ruedi Baumgartnet and Ruedi Hogger (eds.). New Delhi: Sage Publications.
5. FAO (2000), Fishery Country Profile, The Republic of India, FID/CP/IND Rev 5; Food and Agricultural Organisation.
6. Government of India (2000), ftp://ftp.fao.org/docrep/fao/009/ a0692e/ a0692e00.pdf, extracted on August 2010.
7. FAO (1995), Code of Conduct for Responsible Fisheries, Rome. www.fao.org/DOCREP/005/v9878e/v9878e00.htm.
8. Rubinoff, Janet Ahner (1999), Fishing for Status: Impact of Development on Goa's Fisherwomen. *Women's Studies International Forum*, Vol. 22, No. 6, pp. 631-644.
9. Moser, S.C. (2007), In the Long Shadows of Inaction: The Quiet Building of a Climate Protection Movement in the United States. *Global Environmental Politics* 7(2): 124-144.
10. Dujovny, Eial (2009), The Deepest Cut: Political Ecology in the Dredging of a New Sea Mouth in Chilika Lake, Orissa, India.
11. Shepherd, Andrew (2010), Tackling Chronic Poverty, Chronic Poverty Research Centre, December http://www. chronicpoverty. org/uploads/publication_files/Tackling%2 0chronic%20poverty% 20webcopy.pdf.
12. Ganapathy, R. and Viswakumar, M. (2001), Present Status and Future Prospects of Coastal Aquaculture in India. In *Proceedings of the International Workshop on Aquaculture and Environment*, eds. N.R. Menon, B.M. Kurup, and R. Philip, 1–15. Cochin, India: Cochin University of Science and Technology.
13. Okonta, A.A. and Ekelemu, J.K. (2005), A Preliminary Study of Micro-Organisms Associated with Fish Spoilage in Asaba, Southern Nigeria. Proceedings of the 20th Annual Conference of The Fisheries Society of Nigeria (FISON), Port Harcourt, November 14-18. pp. 557-560.
14. Ayuba, V.O. and Omeji, N.O. (2006), Effect of Insect Infestation on the Shelf Life of Smoked Dried Fish. Proceedings of the 21st

Annual Conference of the Fisheries Society of Nigeria (FISON), Calabar, November 13-17, 2006.

15. Dahl, E.K. and Forsgren, A. (1988), Marketing of Fish from Penthakota, Orissa, India. Report of M. FSc. Mission, Fisheries Development Series 28. Goteborg, Swedish National Board of Fisheries.
16. Ayyappan, S. and Jena, J.K. (2000), Coastal Fisheries in Orissa, an Important Economic Activity. In V.K. Sharma, ed. *Environmental Problems of Coastal Areas in India*. New Delhi: Bookwell.
17. DOF (1998), *Handbook on Fisheries Statistics of Orissa, 1996/97*. Bhubaneswar: Directorate of Fisheries, Government of Odisha.
18. MPEDA (1995), Statistics of Marine Product Export, 2a; DOF, 1998: 76.
19. Breman J. (1994), *Wage Hunters and Gatherers: Search for Work in the Urban and Rural Economy of South Gujarat*. pp. xi + 422, ref. 9.
20. Ostrom, E. (1990), *Governing the Commons: The Evolution of Institutions for Collective Action*. Cambridge: Cambridge University Press.
21. Fisher, Thomas, Mahajan, Vijay and Singha, Ashok (1997), *The Forgotten Sector: Non-Farm Employment and Enterprises in Rural India*. New Delhi: Oxford and IBH Pub. Co..
22. Von Pischke, J.D., Adams, D.W. and Donald, G. (eds.) (1983), *Rural Financial Markets in Developing Countries*. Baltimore, MD: John Hopkins University Press.
23. UNDP (1997), Occasional Paper 25–Poverty, Human Development and Financial Services, http://hdr.undp.org/en/reports/global/hdr1997/papers/jd_von_pischke.pdf.
24. Lamberte, Mario B. (1987), "Comparative Bank Study: A Background Paper," PIDS Working Paper Series No. 87-04.
25. Kwack, S.Y. Howe, Hernandez-Cata, H.E., Stevens, G., Berner, R. and Clark, P. 1981. Assessing International Interdependence with a Multi-Country Model. *Journal of Economics*.
26. Ingham, G. (2004), *The Nature of Money*. London: Polity Press.
27. Shah, Mihir, Rao, Rangu and Shankar, P.S. Vijay (2007), *Rural Credit in 20th Century India: An Overview of History and Perspectives*. http://www1.ximb.ac.in/users/fac/ shambu/ sprasad.nsf/

23e5e39594c064ee852564ae004fa010/ e78490ff090249d06525730 c0030abf9/ $FILE/Mihir%20Shah_rural_credit__April_2007__ epw.pdf, retrieved on June 2012.

28. Ramachandran, V.K., and Swaminathan, Madhura (2004b), *Financial Liberalization, Rural Credit and Landless Labour Households: Evidence from a Tamil Nadu Village, 1977, 1985 and 1999.*
29. Samal, Kishore C. and Meher, Shibalal (2003), "Fishing Communities on Chilika Lake Comparative Socio-Economic Study". *Economic and Political Weekly.*
30. Schrader, H. (1993), Moneylending and Rural Indebtedness in Colonial India, Working Paper, University of Bielefeld.
31. Das Gupta (1990), *The Business of Lending: A Typology of Money Lenders* 9-12, http://www.gdrc.org/icm/suppliers/ml.html.

6

Fishermen and Mean Machines

Of all the factors discussed so far, the sub-plot of mechanization is probably the most insidious and virulent. It arrives packaged as a panacea to the ills of marginalization, with the promise of utopian livelihoods and progress, camouflaged in the single mantra—mechanized might is right. The vulnerable traditional fishers are easy prey for the deft manipulation of the multifarious vendors of mechanization. Resource depletion, ecological fallout, iniquitous profit maximization, political exigencies, and non-participatory governance have compounded the crisis of mechanization, in the vulnerable fishery livelihoods of the Chilika Lake. In this chapter, we examine, analyse, elucidate and substantiate these issues with inputs from the fishers of Chilika.

The extent of mechanization is best introduced with estimates of mechanized fishing boats from the FAO. Of the 4.72 million fishing boats available worldwide, a staggering 57 per cent are motorized; with Asia accounting for 72 per cent of motorized boats (FAO, 2012). In India, with the help of UNFAO in 1953, a large number of boats were motorized at a relatively small cost and without affecting the ownership structure and work organization in the fisheries. This was a major factor in creating a conflict free process of new technology adoption and diffusion (FAO, 1958). Apart from fishery-related activities, which use both non-motorized and motorized boats; motorized vehicles include dredgers for maintenance, besides ferries and local boats for tourism; the latter facilitated by the Odisha Tourism Development

Corporation in Chilika. The boat movement indirectly affects the fishery livelihood, threatens traditional fishery practices, and poses environmental threats to the flora and fauna in the lake. In the 21st century, increased influx of boats and high-end technology, far from being a blessing, has increased occupational challenges, livelihood insecurities and societal disruption.

Resource depletion is a globally recognized phenomena, with fishers becoming victims of unregulated practices of fishing, resulting in increasingly low volumes of catch (FAO, 2016). Illegal, unreported and unregulated (IUU) fishing, and related activities threaten national, regional and international efforts to secure long-term sustainable fisheries and promote healthier and more robust ecosystems (Damanaki, 2011). Fuelled by corrupt practices, the recognition of IUU as a "global scourge", prompted its coverage in the 2010 Secretary General's report to the UNGA.[1] The resolution underscored the threat to fisheries and their ecosystem, and the need to intensify fishery monitoring, control and surveillance. Similarly, from the study conducted by M. Binkley (1996), it was concluded that critical reductions in catches of ground fish have led to layoffs or work reductions, changes in work organization, and exploitation of other fisheries, some of which had previously been under-utilized. In the context of this chapter, resource depletion is precipitated by the unbridled power of mechanization, in Chilika in particular, and the world over in general.

Ecological fallout or degradation and climate change have been recognized as the two most critical problems in the fisheries. The other issues which are being highlighted in a number of studies, include unlawful and unregulated fishing practices, leading to destruction of biodiversity. In order to sustain marine fisheries, the Government of India implemented the Marine Fishing Regulation Act 1983 and amended it in 2000. The restriction of fishing for 45 days every year promulgated along the coast line has remained far short of effective implementation (Syafputri, 2014; Drammeh, 2007). Thus, issues like over exploitation of marine resources and under estimation of marine environments, subsequently affected the larger ecosystems and posed a serious challenge to the fisher community (Vivekananda, 2011). The fishery and aquaculture sector now confront a host of other key challenges such as inefficient use of motorized boats, increased contamination of water bodies and ecological degradation surrounding the coastal areas including the consequent impact of climate change. In addition to this, overuse of motorized fishing boats in coastal India, including Chilika is encouraging the processes of social displacement, ecological alteration and other related disruptions (Nayak, 2014).

Iniquitous profit maximization naturally followed the discovery of 'ease of doing business' with mechanized boats; consequently, they superseded the traditional fishery sector, with its much older history. Thus small motorized vessels to big trawlers became part of the post-independence fishery sector in India, along with improved road communication, better storage facilities and higher capital investment. Practitioners of development tried to confront low productivity in a 'sea of infinite fishes' by bringing in mechanized crafts. The profit potential in mechanization was seized by the small, but capitalist group, who immediately invested in the business, funded mechanized boats and improved quality of nets, therefore resulting in increased catch. The financers were ready to advance money to the fishers to tide over expenses of oil, fuel and repairs, in return for the right to sell the fish caught on the boat. It resulted in the fishers losing increasing amounts of their catch and becoming enmeshed with hierarchy of middlemen. Camouflaged and propagated as being socially desirable and wholly undertaken for national interests, the ultra modern fishery sector and its protagonists heightened the profit motive with an intensely predatory 'want of satisfaction"; along with a robust business diversification strategy (Kurien, 1978).

The demarcation of fishery resources prescribed by the Central Cooperative Marketing Society and the "Anchal Adhikari" (the local government officials), have been beset by political exigencies and non-participatory governance. The subsequent shrimp culture conflicts, rental systems, and joint ventures through the Chilika Aquatic Farms Limited (CAFL), resulted in the grass root Chilika Bachao Andolan (The Save Chilika Movement) in the 1990s, which we have discussed in the next chapter. In 1999, through the efforts of the Chilika Matsyajibi Mahasangha, (a fishers' union), the Chilika (Regulation) Bill 2002 came into being to protect the traditional fishers' rights to fishing.

Hence, it is important to look beyond the economic side of fishing activity, ensuring ecological viability in harmony with human well-being. Thus, in spite of the lack of secondary data pertaining to these subjects, an attempt is made to relate macro ecological processes in the life of the traditional fishing people. It includes understanding their perceptions, knowledge and behaviour in relation to increased mechanization in the context of use of nets and boat mechanizations.

Ecological Variation and Boat Automation

Kabibara Radhanath Roy's poem "Chilika Kavya", as stated in the prologue, brought alive the lake's bounty of flora and fauna, through his evocative description of its jungles, rivers, varied bird life, plants, flowers and fish. Kabibara's Chilika, sustains our souls and livelihoods and cannot be lost to ravages of development that does not promote community and ecosystem well-being.

Ecologically, the lake offers fresh, brackish, high salinity and marine zones through its northern, central, southern and outer channel sectors; which attracts migratory birds, tourists and scientists from across the globe, around the year. In addition, the outer channel is generally marine, though it is flooded with fresh water from the lagoon mouth during the monsoon. The lake is influenced by three sub-systems, including the Mahanadi River system, rivers flowing into the lake from the western catchment, and the Bay of Bengal. The lagoon receives freshwater from a series of 52 channels. The Chilika drainage basin, including the lagoon itself, covers an area of over 4,300 kilometres^2 (Das and Samal 1988). The watershed boundaries lie between the water flowing into the Mahanadi and Chilika in the north, while the areas draining into the Bhargavi River make up the north-east watershed; in the west and south-west, the watershed boundary lies between the streams flowing into the Rushikulya River, and those flowing into Chilika

(Ram et al., 1994). The salinity level in the lagoon is thus dictated by river discharges during different seasons, wind action, and the extent of tidal influx carrying marine water from the Bay of Bengal. The seasonality of rainfall also leads to large fluctuations in the size of the lagoon.

Based on the data collected during 2011-12, CDA came up with an ecosystem calculation grade for the lake, highlighting that most of the water quality in these four areas met the desired levels, often leading to a congenial environment for fish and shellfish. On the other hand, the report further highlights that the lake failed standards for total chlorophyll concentrations, that is vital for production of carbohydrates by photosynthesis. Balliarsingh et al. (2014) after undertaking a detailed study revealed that each area exhibited different levels of Petroleum Hydrocarbons Concentration (PHC). Thus, the lake had 50 μg /l, 130 μg/l, 280 μg/l and 330 μg/l in its northern area, southern area, central area and outer channel respectively. The sustainable ecosystem of the lake is threatened by an uncontrolled influx of motorized boats, and overall ineffective management; culminating in increased water pollution, decreasing salinity, siltation, commercial encroachment, and shrinkage of water area, among many other threats. Mechanized boats disrupt the ecosystem through their discharge and propulsion on the lake. Fuels, oils, lubricants and fumes are discharged into the lake affecting the phytoplankton and disturbing the total biological productivity and food web. Turbulence and noise from motorized propulsion increases the negative impact of mechanization on the ecosystem; disturbing the already endangered Irrawaddy Dolphins, and disrupting fish breeding and spawning grounds in the lake. Interaction with fishers reflects that only a mere handful of boatmen have the tourism licenses, among all the boats plying in the lake near Satpada in Puri district. Satapada is one of the most popular tourist destinations in Chilika where dolphins and the sea mouth attract a large number of visitors. This is mainly a winter phenomenon when migratory birds, tourists and their boats, display an exponential rise. Following cyclone Phailin in December 2013, Sahoo et al. (2014) monitored the lake water quality, and highlighted that large-scale quantitative and qualitative variations could alter phytoplankton species composition, leading to food web and lake economy disruption. A report on "visit to Chilika Lake, Orissa" (2008) made by the Planning Commission, Government of India, highlighted the serious impact of the movement of the motorized boats in the lake. The study also reflects the respondents' views on whether mechanization of boats helps fishers in getting more catch and ultimately making more profits. The findings show that although motorized boats greatly facilitate obtaining far larger volumes of catch when compared to non-motorized boats, this is usually attained at the cost of larger environmental degradation. As long as the boat owners make easy money, it incentivizes fishers, and motivates an increasing number of outsiders to enter the fishing business. Being a common pool resource, the Chilika Lake with its free entry and competition, should be a model of economic efficiency, but it seems to promote exactly the opposite desideratum.

The Lake Governance

Post-independence, in 1953, when the fishery sources came under purview of the Government of Odisha, the areas in the lake were rented out by the "Anchal Adhikari" (the local government officials) through public sale to the fishers. In 1959, the Central Cooperative Marketing Society was formed in Balugaon to further work as an apex organization to effectively manage the leasing system. However, till 1988, the cooperative societies could not deliver efficiently in the absence of a clear differentiation of fishing sources, nets and fences. In fact, in 1986, there was a difference between two primary societies on fixing of boundaries for catching shrimp. The fishers in the upper area put long barricades which blocked the overall fish flow to the lower area. Consequently, orders were promulgated by the Government of Odisha in 1988 and 1991; the first prescribing demarcation of the fishery sources and the second separating the capture and culture fisheries in the lake. In actual practice, with neither capture nor culture was clearly delineated; thus, the right to capture remained with the fishers and the right to culture was opened to the non-fishers. The persistent ambiguities resulted in irrational and inequitable distribution of fisheries, countenanced rampant illegal subleasing of fisheries with tenuous mechanisms to control irregularities. Not unnaturally, this policy created more misperception and conflicts than resolutions.

At this point, it would be pertinent to recall the initiation of shrimp culture in Chilika Lake. In the initial phase, the shrimp cultivation was introduced in the outlying landmass of the lake, but widespread and larger areas, were soon transformed into shrimp culture ponds. From 1980 onwards, the trend intensified with central and primary societies, committed to extensive leasing, easing entry of non-fishers and outsiders; the resultant illegal encroachment, chased shrimps for profit with mercenary zeal. Before long, the remaining deeper portions and the entire traditional fishery areas were colonized into massive shrimp culture zones. Lured by the big profit of shrimp culture fishery but unable to afford the requisite capital investment, a number of primary societies took the easy route of leasing or renting out to non-fishers. However, the leasing policy in 1991 revised by the Odisha Government led to a collapse of such rental systems. The policy further brought a differentiation between capture and culture fishing and allowed the non-fishers to get into 'Shrimp Culture' in the Lake. As a result, the highest bidders could access the sale of leases, whereas the traditional fishers' rights to entering and using the lake water were overlooked. Further, it led to the formation of Chilika Aquatic Farms Limited (CAFL), a joint venture of the Government of Odisha and Tata Company in 1991.

Consequently, the fishers perceived the revised leasing policy as a repressive policy against them and the joint venture as a process of privatization against local people's rights. This coalesced into a public issue which led to a protest movement called Chilika Bachao Andolan (The Save Chilika Movement) in the 1990s. In 1999 the Chilika Matsyajibi Mahasangha, initiated a "pulling down of gheries" (barricades for culture fisheries) drive, confronting police and shrimp mafias. Post the given deadline, the fishers themselves started demolishing illegal shrimp farms and during this incident, three fishers died in police firing. Consequently, in 2001, bowing to the pressure from different groups, the Government of Odisha postponed the lease for shrimp culture in Chilika. Subsequently, the Chilika (Regulation) Bill 2002 came into being which protected the traditional fishers' rights to fishing. In 2008 the Matsyajibi Mahasangha, announced stopping of all vehicular movement on the roads of Bhubaneshwar, the capital of Odisha, if the Chilika Bill was tabled in the Assembly by the government. In April 2010, the fishers once again started abolishing shrimp farms; but they stalled it when the government assured them that all the illegal farms would be removed; though, no action was ever taken. Angered by this inaction from the government side, the fishers started the demolition drive of shrimp farms again in mid-January 2011. Concurrently, the Orissa High Court instructed the state government to formulate a comprehensive policy for safeguarding and regulating the fishing practices in the lake water.

Another aspect of the fisheries that deserves attention under governance laws, is the issue of illegal fishing, using illegal fishing nets. This is a global phenomenon that primarily affects developing countries, resulting in loss of resources, income, and taxes from fisheries. The intense competition for the resources among fishers of all scales, can result in declining catch per unit effort, shrinking economic returns from fishing operations, and the over-exploitation of the resources. Zero netting is an added detractor, indiscriminately harvesting even the smallest size of fish, shrimp, crab, with their less than 10 mm net size (zero net, mantle nets and net box). Circumstantial reports highlight the considerable decrease in the abundance of some fish varieties, as a fallout of zero netting. In addition, prodigious amounts of synthetic, non-biodegradable nets also get lost or discarded in the lake water; and remain to haunt the lake as ghost nets and even trapping the unsuspecting fish. Without proper research, not much is known about the magnitude of this issue in the fishery sector; but with anecdotal evidence, it paints a gloomy contemporary picture. At the global level, keeping in mind the unsustainable fishing practices by using zero or drift nets, the United Nations in 1993 banned certain kinds of nets longer than 2.5 kilometres.

Fishing Nets in Chilika

Fishing nets like the boats have been transformed beyond recognition; consequently, their impact on fishing, environment, livelihood, compliance with Orissa Marine Fishing Regulation Act (OMFRA) and prevention of illegal fishing form an important part of our narrative.

Till the early 1980s, or even later in some areas, most artisanal fishing systems in Odisha were dependent on one or two types of locally produced, handmade cotton fishing gear. These were heavy, took time and effort to make and repair, were not very efficient in catching fish and were mostly non-selective. Beach seines and traditional trawl nets are prime examples of such traditional gear. From the mid-1990s, when the decline of catch began to impinge upon returns in a significant manner, necessitating tightening of operations, fishers started using only those nets that ensured at least a minimum return on investment or that targeted more expensive commercial varieties such as shrimp or seer fish.

When synthetic fishing gear was introduced, it revolutionized fishing methods. These nets were available literally 'off-the-shelf', were light and practically invisible once inside the water and could be made very selective. Soon a wide range of synthetic fishing nets of different materials, twine, mesh size, length and width began to be used for a wide range of applications. Synthetic gillnets and trammel nets (called 'disco nets' locally) soon came to replace most of the traditional nets and in a few cases, traditional nets (such as beach seines) were rigged with synthetic material). During data collection, we tried to understand the kind of fishing gear which is principally used in the lake. This could help in deciding whether the extensive use of any fishing gear is blocking the passage of the fish to the lake. It will also help to understand the nature of fishing in the lake and how this can be further strengthened. There are several types of fishing gears operating in Chilika Lagoon, e.g. fixed net (Khanda), gill net, drag net, sieve net, long line, scoop net, and cast net. The fishing population consists of sub-groups such as Kandaras, Khatias, Niary, Nolia, Tiaras, etc. who use different fishing gear and catch different species of fish (Patil, 1970). Khanda (traditional fixed net) is the most popular fishing gear in the lagoon and it is used by more than 49 per cent of respondents. Other major fishing gears are Dubi jal, Menjia jal (both are kinds of gill net) and Zero net (sieve net). Twenty-three per cent respondents feel that Dubi jal is mostly used and 13 per cent feel that zero net is predominantly used. Only 11 per cent feel that Menjia jal is principally used and 4 per cent feel that khepa jal (kind of fishing net) is the most used. These gears are used in Chilika the year-around. The study also reflects an interesting finding on the linkage between fishermen's sustainability to natural disaster and the use of nets in the lake. In the places where the water level is low, most of the fishermen who go into the lake and suddenly come up against a storm, contrary to conventional norms, do not try to move back quickly. Instead, they hold onto the bamboos and remain in situ till it blows over. Most of the time they get injured and lose their nets and boats, but they do survive. However, Mr. Chitta Shankar of Balugaon village had an unusual but reasonable explanation:

> It is not always possible to come back very fast as the whole of Chilika is full of nets, especially where the lake is not deep; you need to cross them very cautiously. It may so happen that your boat will get trapped with those nets, and then, you face even more problems. Therefore, we prefer to stay in the lake for some hours and then come back. During the night it becomes even more dangerous to come back crossing those nets.

It is therefore more important to regulate the fishing nets in Chilika, to avoid any untoward incidents, especially during severe storms.

Perceptions of Legal and Illegal Fishing

In Chilika, the lines between legal and illegal fishing continue to remain ambiguous, in spite of the many regulations framed to protect fishers and their catch. The ambiguity generates non-compliance, with the fishers not following most rules and regulations. Consequently, no measures for fishery resource management have been formulated or practised. To make matters worse, fishing activities in Chilika, are not governed by dedicated and specific acts or regulations. Instead, Chilika comes under the larger and more generalized ambit of Orissa Marine Fishing Regulation Act and Rules (OMFRAR) amended from time to time. Some regulations, pertinent to Chilika, and the target species of Fisheries Resource Management Plan (FRMP), from the OMFRAR are listed.

- Fishing by any method is completely prohibited in the outer channel from July to January; but fishing by net is completely prohibited throughout the year.
- Fishing by any method is prohibited in Palur canal throughout the year.
- Fishing of shrimp juveniles of Bagada below 100 mm size by any method is also prohibited throughout the year in the Chilika Lake.
- Procuring Khainga and Bhekti below 150 mm size through any method is also prohibited throughout the year in the Chilika Lake.
- Use of Zero net or nets with mesh size between 8 mm-10 mm is prohibited in the Chilika Lake throughout the year.

With the increased number of fishermen getting into fishing occupations, and the common property aspect of the Chilika lagoon, it was anticipated that information on illegal fishing, and use of illegal fishing gears, would open interesting leads. This information could profile the fisher community engagement with illegal fishing, and evolve strategies for strengthening their knowledge and bolster environmental sustainability. Through interpersonal interactions a profile of issues such as major fishing gear used by fishermen, their impact and knowledge of illegal fishing has been developed. About 46 per cent of fishers answered that they were aware of illegal fishing regulations, which included fishing gears banned by the government, and areas prohibited for fishing. In contrast, about 54 per cent of fishers said that they had no idea about fishing gears banned by the government and illegal fishing. Not all information regarding their knowledge of illegal practices and fishing was easily obtained. For instance, the fishermen of Arakhakuda village initially claimed total ignorance about the illegal fishing gears. Later, while travelling on a boat with one of their fishermen, we were told that they never received any information from the government about fishing gears banned by the government and illegal fishing. He was also of the opinion that the increasing number of non-fishers, who are involved in illegal shrimp culture and other fishing activities in the lagoon, is a cause of growing concern. Another fisherman raised the issue of information being selectively shared only with those who attend meetings called by the government. By default, a limited number of people get this information and in turn, the attendance for meetings is very low, and the dissemination of information restricted.

We tried to find out whether the fishermen were using illegal fishing gear which can prevent and/or obstruct migration of fish between the sea and the lagoon. At the same time, it may also lead to reduction of a particular fish population from the lake. About 48 per cent of respondents answered that there were many fishers who were using illegal fishing gear in their village. Whereas 23 per cent said that only a few fishers were using illegal gear and 29 per cent answered that fishers were using no illegal gears at all. The illegal fishing activities can be classified into two categories, individual fishers and those managed by PFCS or the village. The illegality might arise from the fishing gear used or the designated areas of the lagoon used for fishing. For instance, zero net, a designated illegal fishing gear, remains a popular gear, along with the shrimp trammel net and shrimp pen culture (gheri culture). On the other hand, villagers or PFCS sublease their fishing grounds already leased by government for shrimp gheris. In addition, many fishers are now autonomous in their choice of fishing gear, while earlier it was dictated by their sub-caste. Some of the respondents revealed that the number of Khanda set in and around the outer channel area, are blocking important fish migration routes. They also informed us about the lease agreement between FISHFED and PFCS, and the limits imposed on fishing gears, such as the size of the Khanda. By regulation, the length of Khanda should be (within 800 ft) and the interval between each Khanda should be more than the stipulated 500m, however no one adheres to it.

Our interactions and deliberations with key stakeholders made us aware of the fact that the rampant use of illegal fishing gears by the fishers is a matter of serious concern not only for the fishermen but also for the government. The fishers trapped by decreasing volumes of catch can hardly stem the rise of unsustainable and illegal practices, even though they comprehend the negative implications of these activities. Fishers admit that use of Alimi nets, zero nets, box nets (different fishing nets capturing small to big size fishes) are not allowed in the outer channel of Chilika. This study also reveals the maximal but unregulated use of the Khanda fishery in the lagoon, is very likely to create problems in future in terms of fish sustainability. It is therefore suggested that the blockage of fish migratory routes by the gear like Khanda should be limited, to ensure the passage of fish and spawners to and from the sea. The lease agreement between FISHFED and PFCS, on the condition of fishing gears, including Khanda should be strictly followed, to ensure increased and sustained fish production.

Consequently, the CDA and Department of Fishery and Animal Husbandry, Government of Odisha should advocate enforcement of the existing Act on Palur Canal and Outer Channel to remove the existing illegal fishing and blockage to protect spawners of target species. A strong regulatory programme to protect, restore, and enhance the estimated spawning grounds of major fishes is also advocated. Introduction of fisheries resource management (fisheries regulation) in terms of regulating the fishing gear is indispensable for sustainable and non-exploitative use of the resources in Chilika Lagoon. However, it is also anticipated that by enforcing any sort of regulations, the income of the fisher community will be reduced in the short term; and therefore, promotion of alternative income activities is also required for introducing fisheries resource management.

While it was revealed that use of illegal fishing gears are common in the lake, it was also important to understand whether the fishermen consider that illegal fishing gear should be stopped or not. This can justify whether the illegal gears can be stopped with the community's participation. The opinions of the fishermen matter, as a society cannot flourish unless all stakeholders including the community contribute. We therefore tried to understand the opinions of the fishermen regarding illegal fishing. This will help to build an action plan for promoting legal fishing practices in the lake, which will ultimately ensure environmental sustainability in the area. Although about 47 per cent of fishers think that use of the illegal fishing gears should be stopped, 34 per cent of them said that these fishers have no alternative solutions, other than using these gears. Similarly, 19 per cent think that the fishermen should decide the choice of their fishing gear.

This clearly indicates that while the fishermen are concerned about the illegal fishing activities, they have no alternative recourse to livelihood, attenuated by the competition they are currently having to combat. The increased market price of shrimp and other fishes also is a reason, which is forcing the fishermen to take up illegal fishing. Our study clearly reflects the duality of the situation in which the fishers are both aware of the negative aspects and yet succumb to illegal fishery activity, consequently jeopardizing the outer channel by the intensive and illegal use of these nets.

Boat Mechanization and Environmental Changes

To begin our discussion, it will be worthwhile to quote at this juncture, the gist of our conversations with NGO officials, CDA officials and locals.

> The first motorized boat was used in Chilika for ferry ghat (Balugaon to Satpada and back) in 1974. Since then the numbers of mechanized boats have increased many fold. This has helped in increasing the fish catch. While interacting with an NGO representative in Krushnaprasad Block, we got to know that the overall motorization of fishing boats is very high (at least 50 per cent) and is moving towards an alarming situation. Similarly, he informed us that there are countless landing centres on Chilika. However, one of the CDA officials informed us that there are around 33 per cent mechanized boats in Chilika and only 18 landing centres are registered with CDA. Motorizations of boats are done in Ganjam in the southern sector, Brahmagiri

in the northern sector and Krushnaprasad in the central sector. He also clarified that the other landing centres are not registered and there are no authentic data available with CDA about this. Seeing the success of mechanization, facilities like storage, processing, supply of fuel and transport were developed further, and the existing facilities were also fully utilized. This made the fishing industry more lucrative, which led to a greater number of fishermen getting into the fishing business.

According to the Department of Fishery (2000) Government of Odisha, there are five categories of fishing systems such as small trawlers, large trawlers, deep-sea trawlers, motorized boats (*Bhutbhuti*) and non-motorized boats. The department initially revealed that a total of 454 motorized and nearly 5,000 traditional boats operate in the lake (DOF, 1998). However, a study undertaken during 2008 by the Chilika Development Authority shows that there were 7,348 fishing boats (2,259 motorized boats and 5,089 non-motorized boats) that operate in the lake. As we present our study, we anticipate that by now the number of boats has increased substantially. Many members of the fishing communities were boat owners (66 per cent), and many of them (27 per cent) had hired crews (CDA, 2009), with only 4 per cent keeping boats on a partnership basis. Out of the total boat owners 37 per cent have mechanized boats, whereas 63 per cent own non-mechanized boats. However, a small number of these boats are used for tourism from about fifteen small boat-docking sites. Our study also reveals that around 57 per cent of sample households owned their own boats whereas 31 per cent of sample households hired crews. There were only a few sample households (about 12 per cent) who keep boats on a partnership basis. This means most of the fishers have either been using the motorized or non-motorized traditional boats for fishing. As per the study made by CDA, the motorized boats constituted 46 per cent of all boats taken into account in Chilika till 2008. In fact, there was an increase of motorized boats (37 per cent) compared to manual boats from 1998 to 2008 in the lake.

During our study, we tried to understand the problems that the fishers confront due to the increasing impact of mechanized boats. Twenty-seven per cent of the households we spoke to told us that motorized boats damage the fishing nets, while about 19 per cent of the respondents conveyed that the mechanized boats violate access and boundary norms of traditional fishing. Among the respondents, 21 per cent expressed concerns about pollution, 4 per cent were worried about over-fishing and 16 per cent spoke of disturbance to fish breeding by the use of mechanized boats. This analysis suggests that the fisher community does not completely approve of the boat automation for their future livelihood.

Many also argue that they do not have any options against mechanization of boats as it instantly increases their incomes in the lake. In this context, as much as 68 per cent respondents perceive that mechanization of boats helped in augmenting the catches, whereas only 24 per cent feel that it did not help. The finding further anticipates that in the future most of the boats in Chilika will be motorized, unless some jurisdictional restriction is strictly enforced. Most of the older and traditional fishers were however, in favour of non-mechanized boats, signalling their priority in earning livelihoods, supported by local resources in a sustained, environment-friendly manner. In fact, as the number of fishing boats increases, the fishery productivity becomes stressed and the fish catches get diminished. In recent years, the traditional and elderly fishers in Chilika face stiff challenges from young fishers, and local government officials who strongly argue for the boat mechanization. Thus clearly, the fishery as a sustainable livelihood for the fishers is under threat, as the objective towards instant market gains for the younger generations by mechanization, can lead to a resource crisis by over-exploitation.

The boats, both traditional and motorized that are increasingly used for fishing, ferrying tourists and transporting the public have been the lifeline for many people in and around the Chilika Lake. While their overall number has substantially increased over the decades, there has also been proportionate increase of the motorized boats contributing to environmental degradation. As we have discussed earlier, the motorized boats compete and conflict with traditional fishing boats in multiple ways; the motorized boats cause over fishing, disrupt fish breeding areas, pollute lake water by leakage of engine oil, damage traditional fishing nets by mechanized vessels and violate access/boundary norms of traditional fishing by commercial ventures. The increasing levels of ecological degradation are attributed by the CDA, to high density of motorized boats in Ganjam in the lake's southern zone, Brahmagiri in its northern zone and Krushnaprasad in the central zone of Odisha. In addition to the plying of mechanized boats, the increased use of different fishing nets and gears has threatened the aquatic species of the Chilika lagoon including the highly endangered Irrawaddy dolphin.[2] However, deaths due to injuries caused by the propellers of motor boats remain a matter of concern, forcing the government in the Forest and Environment Department to order fixing of propeller guards as an interim measure.

The state authorities have not been completely inactive; for instance, during 2009-10 the Chilika Development Authority took an important step to encourage the use of noiseless mechanized boats in the lake. Further, as per the Supreme Court order stated earlier, the building of shrimp culture ponds within the Coastal Regulation Zone and within 1,000 metres of Chilika Lake were banned, except for traditional types of ponds (Supreme Court of India, S. Jagannath vs. Union of India & Others, 1996). However, this could not stop the ever increasing disruptions in the fishery livelihood caused by boat mechanization and commercial ventures in and around the Chilika Lake.

It is vital to state at this point, that over the time, the fishers have also evolved their own management systems. The guidelines and regulations they have framed, have helped in controlling their relations with the water body, managing-fishery resources; in addition to minimizing and regulating fishing related conflicts. Unfortunately, our respondents strongly feel, the government chooses to ignore the significant stakeholder contributions, based on experience, perception, and management. Besides the traditional fishers and government establishments, there are also several other stakeholders such as boat owners' associations, boat federations, fisher cooperatives, women's self-help groups, NGOs and research organizations who have a strong presence around the lake. However, they do not seem to be coming together; this is another aspect that must be taken care of by the government for the efficient and effective management of the lake.

Mechanization and Crisis Over Fishery Livelihood

The lake as a water body is significant for the sustenance of local communities; it not only supports the fishers' community lives, and livelihood, but also sanctifies their cultural space where their traditional and spiritual activities pass from generation to generation. In the past, several studies have drawn significant attention to the small-scale fishers' status, highlighting the fact that livelihood was largely responsible for their poverty-stricken situation (FAO 1995, Rubinoff 1999, Sekhar, 2004).

Our observations and focus group discussions, point towards the increased motorization of boats as the outcome of capitalistic ventures that are gradually alienating the fishers. The original stakeholders are discernibly distanced from owning, controlling, sustaining their capture and production methods for fisheries and aquaculture in the Chilika Lake. The current fishing organizations, in spite of their apparent fishery business acumen, are largely isolated from the earlier community-based dovetailing and brotherhood that ensured safety, security, and sustainable fishing within the lake environment.

Our field observation and focus group discussion also reflected on the economic fallout of mechanization, its high levels of investment, and concomitant unpredictability that would ultimately destabilize the business. While big business entities would try and maintain a particular set of fishers by paying more or by paying in advances, the small-scale fishers in the non-motorized sector would become more dependent. In the lean season, they would be forced to move from boat to boat in search of employment opportunities, which would bring more disparity and inequity among them.[3] Only about 45 per cent of sample households perceived that other stakeholders such as boat owners' associations, fishers' cooperatives, fishers' federation, self-help groups, research organizations and

NGOs have some concerns for fishers and their right to fishing as community livelihoods. Many argue that now the fishing profession is a profitable profession but hereditary concerns on this profession are disrupted by the private and government establishments in and around the lake. To our questions on the crisis of fishery livelihoods, we found different perceptions among different stakeholders. Except traditional fishers, all motorized boat owners hardly comprehended the rapid depletion of fishery resources in Chilika Lake. It seems that everybody was busy converting the fisheries into a total commercial venture. Our field observations and focus group discussions revealed that in the lake region there is a clear technological divergence between the utility of mechanized and non-mechanized boats. Thus, the state-supported "modern" sector involving trawlers and big boats are oriented to catching shrimps and other fishes for the export market. On the other hand, the non-mechanized boats used traditionally by the fishers are left largely to survive on their own as they have over the centuries. While there is visible disgruntlement among the traditional fishers, the newer generation's exposure to modern technology and persisting poverty has led to large-scale mechanization of boats.

Conclusion

This part of the study has broadly covered the issues related to the marginalized fishing communities who are victims of, and vulnerable to, environmental hazards and livelihood disruption brought on by boat mechanization and illegal fishing in the Chilika Lake. The overuse of boats and boat automation is leading to a serious and imminent resource crunch. Except for a small number of rich mechanized boat owners, a large number of fishers, however, do not understand the environmental degradation that is causing severe constraints on their life and livelihood. The fishery and aquaculture's vulnerability to ecological degradation, including degradation brought by traditional fishing communities reveal the crisis and concerns of fishery livelihoods in the lake. The findings of this study strongly argue that periodic monitoring of the water in the lake is vital to address the ecological vulnerability.

The government must ensure that the boat owners obtain licenses for effective and efficient use of the mechanized boats in the lake. All the boats should be periodically checked and issued a pollution control certificate. As the local fishers seem to be imbibing collective conscience for preserving ecological ethics they can be of a great help to the government in controlling excessive mechanization and environmental degradation. For the fisher community, Chilika Lake is an abundant treasure; the instant source of income and livelihood. However, it is quite evident that the fishing population is on a profit-making spree, aided by mechanization; disregarding environmental hazards and its possible effect on the society as a whole. The main conflicts revolve around the boat mechanization and the artisanal fishers' alienation from traditional fishing and sustainable livelihood. Our study also reveals that the damaging impact and subsequent consequences on the fishing community can reach a point of no return. Thus, the issues of mechanization and ecological hazards have their own bearing, on not only livelihood disruptions but also the in-situ nature of common property resources.

NOTES

1. As per the United Nations General Assembly (2011), the issue of IUU fishing was covered. In the UNGA Resolution 65/38, IUU fishing was dealt with at length in Section IV. It called for international cooperation on IUU fishing among states (including through RFBs), besides implementing international instruments and catch documentation schemes (CDSs) and traceability schemes.
2. Taking this into account, the Forest and Environment Department issued an order on November 25, 2005 (No. 20161–EE-37/2004-F&E), notifying "the Irrawady dolphins found in Chilika Lake are endangered".
3. "During the personal interview, the fishermen informed us that the average operation cost of the motorized boats per day is around Rs. 500/- daily and the cost of non-motorized boats is Rs. 100. While this was further cross verified with the boat

owners, it was found that the cost per day for motorized boats on an average is around Rs. 350 and non-motorized boats is Rs. 70. For the motorized boat fuel is the most important cost and for the non-motorized boat it is the cost of the crew. However the boat owner also informed us that this cost varies with the season, with short and long working hours and the increase and decrease of cost towards crew. This also proved that on an average each motorized boat burns around 6 litres of diesel daily and with approximately 2,000 motorized boats, it is expected that they burn 10,000 litres of diesel daily in Chilika, considering not all the motorized boats are not operational every day.

REFERENCES

1. FAO (2016), Illegal, Unreported and Unregulated Fishing, published by Food and Agriculture Organization of the United Nations, Viale delle Terme di Caracalla 00153 Rome, Italy http://www.fao.org/3/a-i6069e.pdf
2. Damanaki, M. (2011), US and Europe Join Forces to Stop Pirate Fishing. In: *Europa* [online]. Brussels. [Cited November 28, 2011]. http://ec.europa.eu/commission_2010-2014/ damanaki/ headlines/press-releases/2011/09/ 20110907_2_en.htm.
3. Binkley, Marian (1996), Nova Scotian Fishing Families Coping with the Fisheries Crisis. In *Anthropological*, Vol. 38, No. 2, pp. 197-220.
4. Kurien, Mathew (1978), *Problems and Prospects of Kuttanadu.*
5. FAO, 2012, *The State of World Fisheries and Aquaculture.*
6. FAO, 1958, *The State of Food and Agriculture.* Rome: Food and Agriculture Orgnisation of the United Nations.
7. Vivekanand, Vincent, Eijsink, G.H. and Svein, J. Horn (2012), Biogas Production From the Brown Seaweed Saccharina Latissima: Thermal Pre-treatment and Co-digestion with Wheat Straw. *Journal of Applied Phycol.*, 24.
8. Nayak, P.K. (2014), The Chilika Lagoon Social-Ecological System: An Historical Analysis. *Ecology and Society* 19(1): 1. http://dx.doi.org/10.5751/ES-05978-190101
9. Syafputri, E. (2014), Almost Half of Illegal Fishing in the World Occurs in Indonesia. https://en.tempo.co/read/

news/2014/07/19/056594269/Almost-Half-of-Illegal-Fishing-in-the-World-Occur-in-Indonesia.

10. Drammeh, O.K.L. (2007), Fisheries and Aquaculture Policy of Liberia. Draft Final Report, p. 36.
11. Das, N.K., and Samal, R.C. (1998), Environmental Survey of Chilika. In S.N. Patro (ed.), *Chilika, the Pride of Our Wetland Beritage* (pp. 96–103). Bhubaneswar: Orissa Environmental Society.
12. Ram, N. Rahul, Rama Rao, N.K.V. and A. Ghosh (1994), *Ramsar Sites of India: Chilika Lake*. New Delhi: WWF India.
13. Sahoo, S., Baliarsingh, S.K., Lotliker, A.A., and Sahu, K.C. (2014), Imprint of Cyclone Phailin on Water Quality of Chilika Lagoon. *Current Science,* 107(9), 1380–1381.
14. Patil, K.C. (1970), Observation on Fishing at the Mouth of the Lake Chilika. In *The Chilika Lake*, 95–102. Cuttack: Directorate of Fisheries, Government of Orissa.
15. DOF. (2000), Rehabilitation Plan for the Super Cyclone Affected Fishermen of Orissa. Cuttack: Directorate of Fisheries, Government of Orissa.
16. DOF (1998), *Handbook on Fisheries Statistics of Orissa, 1996/97*. Bhubaneswar: Directorate of Fisheries, Government of Orissa.
17. CDA (2009), Socio-economic Condition of Fishers in and Around Chilika. The Project Report for Conservation and Wise Use of Natural Resources of the Chilika Lagoon Through Community Participation in India (unpublished).
18. Supreme Court of India (1996), Judgement on S. Jagannath-vs-Union of India and Others, Writ Petition (Civil) No. 561 in the Court of Kuldip Singh and S. Saghir Ahmad on December 12, 1996, 1997.
19. FAO (1995), *Code of Conduct for Responsible Fisheries*. Rome. www.fao.org/DOCREP/ 005/v9878e/ v9878e00.htm.
20. Rubinoff, Janet Ahner (1999), Fishing for Status: Impact of Development on Goa's Fisherwomen. *Women's Studies International Forum*, Vol. 22, No. 6, pp. 631-644.
21. Sekhar Nagothu Udaya (2004), Fisheries in Chilika Lake: How Community Access and Control Impacts their Management. *Journal of Environmental Management* 73.

7

Heritage and Climate: Environmental Concerns

From narratives of men and machines, I now profile the natural environment of Chilika, that is simultaneously a resource and liability, dictating the livelihoods of the fishers, their catch, and, traditional space (sacred or otherwise). It has emerged from the study, that ecological issues including climate induced natural disasters, generate very similar layers of quandary, at all levels. They are characteristically outside the ability of any single entity, organization or state, to manage redressal; and demand collective action, supported by mutual intervention. Further, based on substantive empirical evidence, it is observed that fishery and corresponding livelihoods in Chilika must be well managed, with community participation, so as to ensure long-term sustainability. Thus, effective management of Chilika is dictated by the ability to manage its intrinsic and extrinsic factors. In this context, the dual aspects of ecological wealth and propensity to natural disasters must be counterbalanced to sustain fishing livelihoods, specific to the region. A deliberation on these issues, paves the way for a more empathetic assessment of the daily struggles and strife of fishing communities in and around the Chilika. In this chapter we discuss the fishers' livelihood in the context of environmental degradations, natural disasters and the larger environmental concerns representing the Chilika region. Beginning with its position in the Bay of Bengal, to the impact of monsoons and climate change; siltation and salinity; change in forest cover, shoreline and demarcations; besides bird populations, tourists and poaching have all been brought together, in this penultimate chapter on the environmental issues critical to Chilika.

Environmentalism is not simply concerned with nature but with the sustainable use of nature (Guha and Gadgil, 1992, 1995, Guha and Alier, 1998). In any water body, climate, food availability and quality, suitability of shelter, and the presence of predators (including humans) affect the fish populations.[1] In short, any human activity that affects water quality and habitation has the potential to directly affect the fish population and overall aquatic system and its structure. Chilika has been under constant threat from both natural as well as manmade problems. The socio-economic deterioration of a large chunk of the fishing community, traditionally dependent on the lake, has been deeply impacted by its overall ecological degradation. The geomorphology, water quality and biological productivity of the lake have undergone significant changes over time, under the influence of natural events and anthropogenic interventions (Sahu et al., 2014). Several researchers have established that the natural resources in Chilika are under a great threat (Pattanaik, 2006; Mishra and Griffin, 2010). The non-implementation of the Supreme Court's order to abolish the shrimp gheris, is a cardinal cause of degradation of Chilika; as powerful players with vested interests of intensification of shrimp farming activities in Chilika, ensure no laws are strictly enforced.

Chilika, By Virtue of Its Position

The Chilika lagoon in Odisha situated along the Bay of Bengal, by virtue of its sub-tropical coastal position, lies within the six major cyclone-prone regions of the world. Chilika is not only impacted by the annual cyclones that strike the coastal regions of India (Shanmuga Sundaram et al., 2000) but is also prone to floods, storm surges and tsunamis. The Bay of Bengal has witnessed significant changes in the conversion rate of storms, in the recent past. Thus 42 per cent (10 of 24) of the storms formed between 1996-2010, were converted into tropical storm strength or higher compared to 30 per cent (8 of 27) between 1981-95. The formation rate of strong tropical cyclones was observed to be higher in the Northern Indian Ocean compared to the Western Indian Ocean. This was attributed to global warming induced by increase in Surface Sea Temperatures (SST) as postulated by Balaguru et al. (2014, pp. 3600). With the greatly improved prediction and warning systems available today, it is vital to communicate them to the fishers, who venture into the lake and sea under all circumstances. In addition, it is essential to differentiate between risks on and off the water (i.e. inland vs. lagoon).These factors take on added significance as the fishers venture deep into the lake, the mouth, or even the sea, and confront a heightened degree of exposure to water-related hazards.

Chilika is also impacted by Odisha's densely populated coastal plains, that are the sedimentary deposits of its river systems. With a major portion of Odisha's coastal riverine system draining into Chilika, the lagoon bears the brunt of the heavy loads of silt, flooding and breached embankments.[2] Significantly, when Odisha experienced a 'Super Cyclone' in 1999, the worst catastrophic cyclone in a hundred years; the multilevel (Government, CSOs and communities), preparedness to meet a mega-calamity was a magnificent shift in reduction in life loss and disaster management. On the negative side, no mechanisms were in place, to prevent the monumental economic loss.

Impact of Monsoons

Odisha has a typical limited monsoon period spread over July to September which witnesses more than 80 per cent of the total rainfall in the state, reminiscent of the other states in the eastern belt of India OSDMA (2016). The monsoons are central to multiple concerns of the fishers. For example, the monsoon is the time, when the excessive rainfall increases the water level in the Mahanadi river system, which gradually pumps in fresh water into the lake and pushes away saline water towards the sea. Simultaneously, wind patterns and directions change, reducing fish yield and fishing actives. Unpredictable storms, tidal surges, rainfall and raging winds have become inherent to the monsoon and the lore of fishers. Thus, based on their traditional experience and practices, certain segments of fishers, prefer to opt out of fishing; and engage in household and/or agricultural activities during this period.

In terms of evidence for the above arguments, the study brought out certain important facts and trends cardinal to the monsoons. Forty-four per cent of the respondents feared the spread of diseases (malaria, cholera, typhoid, diarrhoea, jaundice and hepatitis A), 34 per cent marked decrease in fish catch during the monsoon, while 21 per cent of them cited difficulties in venturing into the lake, because of wild winds and tidal surges. In spite of variations in responses from the fishers, a clear pattern of monsoon-related concerns emerged. Beyond the obvious negative aspects, these concerns should also be addressed in the context of life and livelihood, which require holistic management. Erratic weather patterns, add one more dimension that many in the fisher community, may not be able to afford or endure, with the extant practices of survival and protection. Piecemeal solutions would be a waste of resources; therefore, redressal through multi-layered and multipronged processes of planning and action are the need of the hour. Disease surveillance and prevention targeting the occupational and situational vulnerability of the fishers is a vital monsoon strategy; essential for the state as well as other stakeholders working towards substantive changes in the life of the fishers.

Siltation, Loss of Forest Cover

For many years Chilika has been facing the problem of siltation caused by sedimentation. The sediments are flushed into Chilika through soil erosion from bald hills, eroded lands, and degraded forestlands of catchment areas of the different rivers and rivulets. During the interactions, the fishermen informed us that there are ten rivers and their branch rivulets draining into the lake and feeding the wetland with flood water during the rainy season. These rivers and rivulets carry a massive quantity of silt to the lake. Sixty-six per cent respondents feel that siltation is a major issue, which is affecting Chilika's ecology, whereas 30 per cent replied that it is not an issue. At the same time only 4 per cent informed us that they have no idea about this and that siltation really does not matter. Our investigations reflect that siltation in the lake has reached a point of saturation, necessitating in-depth studies to identify the critical issues.[3]

Siltation has caused three noticeable ecological effects, namely shrinkage of the lake area, decline in the lake's depth and weed growth. Similarly, siltation has a major impact on the decrease of fish catch. Chilika is also affected by illegal encroachment of its shoreline, which not only affects the livelihood of poor local fishers, but also has an adverse effect on fish spawning and mullet fishery. All these issues have further been studied during the research and are interlinked with one another.

Siltation is also responsible for decrease in salinity and significant increase in nutrients, causing weeds to spread approximately over one-fourth of the lake. Efforts should be made to remove aquatic weeds, particularly from the western and northern shore of Chilika, and abolish shrimp gheris, which disrupt the tidal flushing and free movement of fishing boats in the lake. Due care must also be taken to reduce this load by the watershed development programme in the catchment areas; besides, extensive plantation should be carried out on the banks of rivers to check soil erosion and silting of Chilika basin. The lake can further be restored by integrated watershed management, with active participation of local community and non-governmental organizations, on a micro watershed basis. The improvements of socio-economic conditions of the local communities, shared decision-making, and improvement of communication network, wetland research and training centre are vital for any measure of success.

During our discussion with the Forest Officials, we were informed that the Department has taken major steps in the past years to preserve as well as generate new forests[4]. Number of awareness generation programmes have also been undertaken on this subject. However due to constant livelihood challenges, the people's dependence on natural forests are also very high. For example, during a focus group discussion, some of the older fishermen informed us that, there were large forest areas both on the coastal side and around the lake, which were covered by casuarina (near the sea shore), eucalyptus and cashew that continued to be cut down by the locals to meet their fuel requirements. Fisher people also use the bamboo to make fishing tools. The Forest Department not taking any measures against these activities is attributed to a range of reasons from avoiding conflicts to operational challenges. Conflict arises due to the inherent nature of community dependence and use of timber and non-timber forest produce for their life and livelihood; which fuels conflicts between the officials preventing such action and the communities. Operational challenges further evolved around limited manpower strength to ensure vigilance to such a vast area.

Similarly, farmlands, mostly paddy fields, are spread all around and irrigation water with pesticides drain straight into the lake, causing potential harm to a variety of aquatic life, including fish. The spread of shrimp culture ponds and shrimp gheris since the mid-1980s all around Chilika has contributed to spread of nutrients facilitating weed growth, which in turn act as a breeding ground for mosquitoes. Therefore, inter-departmental[5] collaboration needs to be strengthened for planning and implementing environmental protection schemes. This will also lead to fishery resource management for improving livelihood of fishers.

The lake's north shoreline comes under Khurda district, while the western coast is part of Ganjam district. The highly productive and sensitive shoreline ecosystem of the lake is altered due to artificial enclosure. Forty-seven per cent respondents feel that demarcation of the shoreline will help, whereas 34 per cent respondents feel that demarcation of the shoreline will not help in protecting the lake's ecology. The study thus proves that the lake's fishery is adversely affected due to destructive fishery practices like encroachment of shoreline for shrimp culture.

The integration of remote sensing techniques with geographical information system (GIS) has proven to be an extremely useful approach for studies in shoreline changes; therefore similar applications using the latest technology can be considered. In employing this technology, care would have to be taken, that decision-making does not get centralized in the hands of agencies and governmental departments, which have use and expertise over it (Gururani and Vandergeest, 2014).

Loss of Avifauna

Ecological change of Chilika is affecting the avifauna of the lake. More than a million birds congregate every winter, but the numbers are falling because of weed growth near Nalabana–the bird sanctuary and the decline in the quantity and variety of fishes on which these birds depend. As previously discussed the decline in salinity, due to siltation and sedimentation is contributing to excessive weed growth in the lake and demands immediate attention. De-forestation and bird poaching in the Chilika basin are also the contributing factors. The decline in avifauna has led to a substantial decline in eco-tourism and fall in arrival of international tourists, resulting in economic loss for the locals who are dependent upon tourism economy.

Despite several security measures, poaching is the biggest threat to the migratory birds here, posing major challenges for wildlife officials (Kadekodi and Gulati, 1999, Ghosh and Pattnaik, 2006). Poaching is also a fall out of livelihood threats, as it is an easy source of income. We therefore tried to ascertain whether illegal bird poaching is still an issue in Chilika. The mandate based on respondents' response was divided. While 58 per cent respondents feel that illegal poaching still continued, the others did not agree. In 1997, the Prevention of Poaching of Waterfowl in Chilika Lake with Community Participation Programme was launched which veered towards the menace of large-scale poaching of migratory and non-migratory waterfowl in Chilika Lake. The intervention strategy focused on discussing problems and strategies with local villagers, comprehending the modus operandi of poachers, and convincing the poachers to give up killing. Another strategy was to engage them in bird protection committees where the local Forester of the State Forest Department was the chairperson. All these interventions primarily led by an NGO called 'Wild Orissa', were effective in reducing incidents of poaching (Pathak and Kothari, 2006).

Conversely, one must be equally vigilant about the false and baseless criminalization of local communities on the grounds that they are poachers. The government thus needs to address the issue of illegal bird hunting by working with local NGOs and CBOs and through the Forest Department of Odisha. It is suggested that in addition to 'Bird Protection Committees', which have assisted in developing alternative economic activities through soft loans; a number of small "bird protect committees" should be formed in a cooperative model. This will require specialized knowledge, which currently may not be easily available in CDA. With reference to bird populations, Sharma and Agnimitra (2015), indicate a growing social concern over poaching over the last century, aggravated by the enormous difficulty of scientifically reliable estimates and causes of decline in the population of migratory birds. In such situations, the scientists and administrators have historically found it easy to accuse native hunting-gathering groups of poaching. Poaching invariably serves to undermine losses through habitat destruction, be it at the hands of large-scale development, or rapid urbanization.

Climate Change

While various exposures to climatic changes in Chilika lagoon have been reviewed through secondary data and research papers; the perceptions of the local population have been inadequately cited or ignored; creating a significant lacuna in understanding their behaviour and responsive actions (Grothmann and Patt, 2005). It is also evident that the fishery livelihoods in Chilika are impacted by the changing environment and variation of aquatic ecologies. Beyond the threat of climate change to natural resources, climate induced changes are significantly affecting resource flows and thus challenging viability of livelihoods. Even though exploration of the direct causal linkage between climate change and disasters continues to be debated, it is fairly well established that climate-induced natural disasters are rapidly increasing (Thomas and López, 2015). Global estimations clearly establish that the increasing pace of climate change will result in extreme weather events and in increasing frequencies. Climate change poses significant risks to the livelihoods, culture and health of millions of people (Barnett, 2003). Ecological and climatic changes are becoming more frequent, resulting in devastation to livelihood, especially the poor living in precarious environments (Ogata and Sen, 2003). In this respect, human beings are becoming increasingly vulnerable.

People living in the coastal areas, by nature of their life and occupation, have lived for generations close to the water for their livelihood. The fishers have always been considered the most vulnerable group, in relation to the onset of any natural disaster. The marginalized and the economically disadvantaged groups are generally the worst affected by natural disasters as they live in particularly vulnerable areas, with very restricted access to aid and support after the disaster.

Defining Hazards and Mitigating Risks

Contrary to popular perception, small-scale disasters are far more damaging than the large-scale ones with regard to fishers, which go unnoticed in mainstream media. It is the fishermen by their occupational exposure to natural disaster events who bear the brunt of extreme weather conditions and thus are among the most vulnerable groups in the Chilika lagoon. The fishers in Chilika have different practices and patterns of fishing based on caste. While some fish within the vicinity, some also sail deep into the sea or to the mouth of the lagoon. In doing so, safety becomes their last priority, as they hardly carry any safety or response equipment, which amplifies their vulnerability,[6] even to small-scale hazards,[7] either natural or man-made. Therefore, for this study, understanding the nature of hazards that the fishers experience while fishing was an important consideration. The study reflects that in terms of frequency, the majority (43.9 per cent) experience high speed wind as they go for fishing, followed by heavy rain (3.3 per cent), flood (1.1 per cent), and high tide (0.4 per cent). Most of the respondents (70 per cent), highlighted that they faced high speed winds along with heavy rain, which affected them the most. Additionally, the trends in temperature rise were seen as posing enormous hardships to the fishers. The fishers felt that while these hazards were more seasonal, their frequency in the recent past has certainly increased.

The hazards are far more likely to scale up from calamitous into full blown disasters, given their extant poverty, defined by the increased competition over scarce resources and deteriorating alternate livelihood options. It is for such reasons and more that Kesavan and Swaminathan, (2006) took note of the vicious cycle between environmental degradation, poverty and climate-change-related natural disasters. Though over the years, progress in understanding implications, inter-relations and integrations of these factors is visible among the responsible stakeholders (both government and greater civil society organizations including academia), the modalities for addressing the issues are yet to be explored fully.

The dynamics between increasing poverty, exposure to stress and further increase in vulnerability have been explored through different approaches. Yet, the sustainable solution to the issues and challenges are far from being comprehensively achieved. A significant paradigm shift in approach to fisheries and their management is visible in the newly accorded Fisheries Policy of Odisha, 2015; according recognition to natural calamities as a key challenge towards the development of fisheries. It has prioritized fisherman's safety, through committed provisions for their personal safety by improving access to early warning, life-saving equipment; besides inter-institutional linkages between responsible agencies. However, gauging safety and livelihood from different perspectives has given rise to a serious disconnect. Ad hoc evaluations of fisher and disaster risks are unlikely to yield sustainable solutions, to counter repeated and current disaster vulnerability. There is a pressing need to concurrently integrate safety and livelihood options with extant socio-economic challenges.

The environmental degradation in Chilika poses a great danger to the fishermen's own survival. It should be realized, sooner than later, that conservation and improvement of the environment are vital for the survival and well-being of fishermen and the entire community dependent on Chilika. CDA along with several NGOs and international organizations have carried out massive awareness programmes in the fishing villages of Chilika. It is encouraging, 66 per cent of respondents feel that awareness programmes help in managing environment.

Article 48 A of the Constitution of India says that "the state shall endeavour to protect and improve the environment and to safeguard the forests and wildlife of the country." List III of the Concurrent List of Schedule VII also provides power to both state legislature and Indian Parliament on various aspects related to environment. Article 21 of the Indian Constitution provides Right to Life which is interpreted by the Supreme Court in a number of cases. The Water Prevention and Control of Pollution Act, 1974, Municipal Solid Wastes (Management and Handling) Rules, 2003; are some of the major acts which regulate the huge water bodies. Similarly, the Ramsar Convention on Wetlands (the first intergovernmental treaty between nations, signed in Ramsar, Iran), entered into force in 1975, is an intergovernmental treaty; delineating the framework for national action, and international cooperation, for conservation and wise use of wetlands. About 156 countries and 1676 sites are designated under this convention as Ramsar sites of international importance. The definition of wise use is given in Article 3.1 of the convention, i.e. "Wise use of wetlands is the maintenance of the ecological character, achieved through the implementation of ecosystem approaches, within the context of sustainable development." According to this convention, every contracting party should adopt national wetland policies, and develop programmes such as research, training, public awareness; review existing legislations and institutional measures for dealing with conservation of wetlands. The Ramsar Secretariat also assists the contracting parties by providing them expertise, guidelines for wise use of wetlands, conservation measures, and funding for implementation of such measures.

There is a lack of knowledge and awareness on the part of fishermen about the legal provisions for environment protection which ultimately leads to environmental degradation. The study reflects that only 14 per cent respondents have basic knowledge about the legal provisions on environment protection. The focus group discussions also revealed that the environment of the lake gets affected by the tourists. The local CDA Office viewed that the local administration keeps a close eye on the tourists and ensures environmental safety even during the high tourist season. However, some of the boat operators informed us that the tourists pollute the lake by throwing harmful substances into the lake such as polythene bags and allied wastes. The problems that tourism might pose for ecological conservation, is differently pointed out by Stonich (1998) and Juarez (2002). They argue that tourism has proven to be damaging for the environment, and indigenous understandings of nature alike. This is difficult to manage, unless the local authorities create awareness of the harmful effects of these acts, and implement laws prohibiting pollution of the lake. Contribution of the local people is extremely important for managing environment safety. Concurrently, the government should enable a conducive environment to local community organizations in managing the local fishers, while helping enforcement of environmental rules and regulations. The state must make a transparent and conscious assessment of the tradeoffs between efficiency and equity, in matters of environmental policy, given the rising volume of detrimental environmental fallouts in and around Chilika. Simultaneously, measuring the effectiveness and gaps in contemporary environment conservation efforts, are vital, in and around Chilika. Such assessments will foster a clear understanding of the tradeoffs, enabling concerted stakeholder action, based on learning experiences from past initiatives.

Perceptions

To question, the nature of that evidence which affirms environmental change in Chilika and impact on the fishery livelihood, beyond the present testimony of our senses, or the academic records, has become mandatory. Without scientific research, we are mostly dependent on the narrations, and existing knowledge of the locals, to comprehend the environmental impact. During the process each of the respondents further substantiated their views and observations with an example or some evidence. For example if you ask a fisher, why he believes that conditions have changed, he would cite signs like the delayed arrival of migratory birds or that availability of certain fishes, have become more sporadic and so on. My inference from these narratives is the undeniable validity of cause and effect of the people on ground. Thus 65 per cent of respondents feel that the overall environment of Chilika is affected during the past few years, whereas 30 per cent feel that the environment has not been affected. During interaction with the CDA officials, it was informed that due to siltation in the north-east and north-west sectors of the lake, the average depth of Chilika had substantially reduced. Previously, much higher numbers of birds visited the lake during winter such as the Yakut from Siberia, the Pochard Duck from the Baikal regions, and certain varieties of teal from the Eurasian belt. The present decline in numbers may be attributed to reduction in landing grounds.

Conservation of environment in Chilika is not only the responsibility of one organization, but is equally pertinent to the various other state departments. As per the Forest Department's Resolution No. EE-82/91.1619/F&E., dated January 24, 1992, the objectives of other departments in resource management is "to survey, plan and prepare the project proposal for Integrated Resource Management for all round development in and around Chilika" but not implementation. Various national and international agencies including Government of India (MoEF), academic institutes, UN agencies, international agencies like Wetlands International and many others are keeping track on the environment of Chilika. The lake was added to the Ramsar List of Wetlands of International Importance in 1981. These measures have not mitigated the threatened ecological balance of Chilika or the simultaneous depletion of its peripheral area. Our findings also reflect that only 11.8 per cent believe that proper steps are being taken to protect the Chilika environment. This proves that the efforts made by the state for the conservation and protection of the Chilika environment are still insufficient. This is generating the biggest threat to life, and livelihood of not only the fishermen living on the Chilika but also the people inhabiting the region.

Conservation of natural resources or their sustainable use has often been regarded as characteristic of traditional Asian cultures (Kalland and Persoon, 1998; Berkes, 1999; Ramanujam and Kadamban, D. 2001, Shastri et al., 2007; Kothari, 2007). The environmental degradation has destroyed the production systems of peasants, fisher-folk, hunters and gatherers in many parts of Asia in multifold ways; beyond immediate quality of their daily lives, their very production has been significantly affected (Omvedt, 1993). The fate of hunter gathering communities, especially those living around habitats scientifically designated wildlife areas, is particularly precarious. Chilika is experiencing rapid decline of resources, which exerts great pressure on the fishing community, whose sole source of livelihood is the lake. In this context, it is useful to analyse the state's claim of environmental conservation activities in Chilika.

Conclusion

Adverse impacts of changes in the climate and degradation of macro and micro-environment, particularly issues like salt intrusion, and increased sediment depositions in deltaic regions have become increasingly visible in coastal areas of India; notably impacting the balance between sea and freshwater in Odisha. In ecologically sensitive areas like the Chilika, these factors have magnified the traditional woes of flooding, water logging, etc. and adversely impacted lives and livelihood of fishers who are traditionally dependent on the lake ecology.

The dynamics between increasing poverty, exposure to stress and further increase in vulnerability have been explored over time and through different approaches. Yet, the sustainable solution to the issues and challenges were far from being comprehensively achieved. A significant paradigm shift in approach to fisheries and their management is visible in the newly accorded Fisheries Policy of Odisha, 2015 which recognizes natural calamities as a key challenge towards the development of fisheries. It has given priority to the safety of the fishermen by committing provisions for their personal safety through improving access to early warning, life-saving equipment, and by creating and using institutional linkages between responsible agencies. However, there seems to be a missing link between safety and livelihood as these are seen from different perspectives. Looking at disaster risks for the fishers and fishing from different and ad hoc perspectives may not result in sustainable solutions, due to the contemporaneous push of vulnerability by repeated disasters. There is a pressing need to look at safety and livelihood from an integrated lens to deal with other existing challenges which are mostly socio-economic in nature.

NOTES

1. Water quality, turbidity, and the presence of pollutants can also affect fish reproductive success, which affects fish populations. Species composition is usually determined by a number of factors including water quality. Fishing has a large impact on fish population and thus, it has to be regulated to try and maintain a balanced fishery.
2. Though a large part of the state comes under Earthquake Risk Zone-II (Low Damage Risk Zone), the Brahmani Mahanadi graben and their deltaic areas come under Earthquake Risk Zone-III (Moderate Damage Risk Zone) covering 43 out of the 103 urban local bodies of the state.
3. Due to excessive soil erosion the water layer of the villages are affected and the water layer of wells, tanks, ponds have decreased and dried. The rain induced top soil erosion has disturbed the bio-chemic nutrition of the land. Hence, the productivity of the land for the last few years has been reduced drastically. Due to decrease in surface moisture, the water layer is rapidly decreasing. Degradation of green forest and soil erosion caused draught like situation in these villages. In addition, to soil erosion and incapability of the lands to conserve moisture is a major problematic area. The grazing land and forestland have continuously decreased over decades; consequently, the villagers face lack of firewood and grazing lands.

4. Divisional Forest Officer, Chilika Wild Life Division, Balugaon, Khurda have been engaged in providing a number of services including Protection of Forests & Wildlife, Creation of New Forests, Eco-Development activities and Soil Conservation in and around Chilika.
5. Department of Fisheries & Animal Resource Development and the Environment and Agriculture Department.
6. Vulnerability has been interpreted as a cumulative consequence of physical, social, economic and environmental processes that limits an individual/communities ability to cope with an emergency situation.
7. Hazards are used to explain potential phenomenon or event that can cause destruction/damage to life, property or environment.

REFERENCES

1. Barnett, J. (2003), Security and Climate Change. *Glob Environ Change* 13:7–17 doi: 10.1016/S0959-3780(02)00080-8.
2. Ogata S. and Sen, A. (2003), *Human Security Now, Commission of Human Security*. New York: Commission on Human Security.
3. Shanmuga, Sundaram, J., Arunachalam, S., Gomathinayagam, S. and Lakshnaman, P. (2000), Cyclone Damage to Buildings and Structures: A Case Study. J. *Wind Eng.* 84:369–380 doi: 10.1016/S0167-6105(99)00114-2.
4. Grothmann, T. and Patt, A. (2005), Adaptive Capacity and Human Cognition: The Process of Individual Adaptation to Climate Change. *Glob Environ Change* 15:199–213 doi: 10.1016/j.gloenvcha.2005.01.002.
5. Thomas, Vinod and López, Ramón (2015), Global Increase in Climate-Related Disasters. Asian Development Bank Economics. Working Paper Series No. 466. November 2015 Available at SSRN: https://ssrn.com/abstract=2709331 or http://dx.doi.org/10.2139/ssrn.2709331.
6. Gadgil, Madhav and Guha, Ramchandra (1992), *This Fissured Land: An Ecological History of India*. Delhi: Oxford University Press.
7. Gadgil, Madhav and Guha, Ramchandra (1995), *Ecology and*

Equity: The Use and Abuse of Nature in Contemporary India. New Delhi: Penguin.

8. Guha, Ramchandra and Martinez-Alier, Juan (1998), *Varieties of Environmentalism: Essays North and South*. Delhi: Oxford University Press.
9. Pattanaik, Sarmistha (2006), Commercialization of Shrimp Trade, Environment and Rural Poverty: A Socio-Ecological Exploration in Coastal Orissa, Final Revised Version of the Research Paper for IEG Working Paper Series, No. E/274/2006 Presented at the Workshop on 'Trade, Environment and Rural Poverty' Institute of Economic Growth, University of Delhi, Delhi. (August 18-19, 2006).
10. Mishra, S.R. and Griffin, A.L. (2010), Encroachment: A Threat to Resource Sustainability in Chilika Lake, India. *Applied Geography*, ISSN: 0143-6228, Vol. 30, Issue: 3, pp. 448-459.
11. Chilika Development Authority - http://www.chilika.com/environments.php.
12. Kalland, Arne and Persoon, Gerard (eds.) (1998), *Environmental Movements in Asia.* Surrey: Curzon.
13. Berkes, F. and Folke, C. (1998), "Linking Social and Ecological System for Resilience and Sustainability". In *Linking Social and Ecological System: Management Practices and Social Mechanisms for Building Resilience*, F. Berkes and C. Folke (eds.). Cambridge: Cambridge University Press.
14. Ramanujam, M. and Kadamban, D. (2001), Biodiversity and Conservation 10: 1203. https://doi.org/10.1023/A:1016637627623.
15. Shastri, H., Barik, B., Ghosh, S., Venkataraman, C. and Sadavarte, P. (2017). Flip Flop of Day Night and Summer-Winter Surface Urban Heat Island Intensity in India. *Scientific Reports*, 7. doi:10.1038/srep40178.
16. Kothari, Ashish (September 2007), "Traditional Knowledge and Sustainable Development." International Institute for Sustainable Development. https://www.iisd.org/pdf/2007/ igsd_traditional_knowledge.pdf.
17. Omvedt, Gail (1993), *Reinventing Revolution: New Social Movements and the Socialist Tradition in India*. New York: An East Gate Book.

18. Stonich, Susan, (1998), 'Violence, Environment, and the Blue Revolution'. Paper Written for the Workshop on Violence and the Environment. Institute of International Studies, Uniersity of California, Berkeley, California, September 24-26.
19. Juarez, A.M. (2002), Ecological Degradation, Global Tourism, and Inequality: Maya Interpretations of the Changing Environment in Quintana Roo, Mexico. *Human Organization* 61, pp. 113–124.
20. Gopal, K. Kadekodi and S.C. Gulati (1999), Root Causes of Biodiversity Losses in Chilika Lake, CMDR Monograph Series No. 26 accessed via http://cmdr.ac.in/editor_v51/assets/mono-26.pdf.
21. Ghosh, A.K. and Pattanaik, A.K. (2006), Chilika Lagoon, Experience and Lessons Learned Brief, accessed via http://www.worldlakes.org/uploads/08_chilika_lagoon_27february2006.pdf.
22. Pathak, N. and Kothari, A. (2006), Protected Areas, Community Based Conservation, and Decentralization: Lessons from India. A Report Prepared for the Ecosystems, Protected Areas, and People Project (EPP) of the IUCN World Commission on Protected Areas.
23. Sharma, V. and Agnimitra, N. (2015), Making and Unmaking the Endangered in India (1880-Present): Understanding Animal-Criminal Processes. *Conservat Soc.* [serial online] 2015 [cited February 9, 2018]; 13:105-18. Available from: http://www.conservationandsociety.org/text.asp?2015/13/1/105/161229.
24. Kesavan, P.C. and Swaminathan, M.S. (2006), Managing Extreme Natural Disasters in Coastal Areas, Phil. Trans. R. Soc. A 2006 364 2191-2216; DOI: 10.1098/rsta.2006.1822. Published August 15, 2006.
25. Balaguru, K., Taraphdar, S., Leung, L.R. and Foltz, G.R. (2014), Increase in the Intensity of Post-monsoon Bay of Bengal Tropical Cyclones. *Geophys.* Res. Lett., 41, 3594–3601, doi:10.1002/2014GL060197.
26. OSDMA (2016), Odisha State Disaster Management Plan. Prepared by Odisha State Disaster Management Authority, Government of Odisha.

27. Gururani, S. and Vandergeest, P. (2014), New Frontiers of Ecological Knowledge: Co-producing Knowledge and Governance in Asia. *Conservat Soc* 2014;12:343-51.
28. Sahu, B.K., Pati, P. and Panigrahy, R.C. (2014), Environmental Conditions of Chilika Lake During Pre and Post Hydrological Intervention: An Overview. *Journal of Coastal Conservation.*
29. Pattanaik, Sarmistha (2007), "Conservation of Environment and Protection of Marginalized Fishing Communities of Lake Chilika in Orissa, India". *Journal of Human Ecology,* Vol. 22, No. 4. December, pp. 291-302.

8
Chilika Moves On

After looking at intrinsic socio-economic vulnerabilities and current livelihood as two central dimensions of poverty, this book discusses some of the long-standing issues related to fishery livelihood in Chilika. Through strong empirical evidence, I have argued that the issue of current livelihood among the traditional fisher communities cannot be condensed to a simple correlation between income poverty, without looking at the more complex and intricate issues like property rights, long-term environmental issues and large-scale modernization. Consequently, the current arguments made in this book are worth taking into consideration; though, more exhaustive analyses are certainly required. The latter must account for the diversity of fishing-related livelihoods in Chilika and enumerate the complexity of causes of poverty, both inside and outside the sectors. The book particularly highlights how deep-rooted poverty among the fishers in Chilika often is, beyond the generic cause and reason, it can be attributed to a wide range of socio-institutional factors, other than income. Thus, landownership, debt, access to health, educational status and financial capital, and marginalization from political decision-making, also become major contributory factors. The empirical information used in this book, including narration of several stories by the fishers, highlights the need for having a more systematic and well planned state support system, to ensure reduced vulnerability among the fishing community.

Chilika, its fish, the fishermen, the fishing operation, and the social practices, are all dovetailed and integral to the ecological balance. In the current situation, not everyone has access to new technology. At the same time increased capital requirements implies that the fish trader or the boat owner may not actually be fishermen, but just personnel involved in business. The introduction of new technologies has made significant contributions to overfishing, with certain encounters with poverty leading to collective movements. At the same time changes in the ownership and management of CPRs has led to increased inter-household inequality. To examine the effects at the intra-household level, it is important to consider the use of CPRs by different household members, and the ways in which their access to these resources is transformed.

Although the need for alternative livelihood, is widely recognized in view of the nature of common property, and efforts are made to address the issue in a meaningful way, it falls far short of the actual requirements. While reflection on multifaceted vulnerability of fishers in the light of property rights and increased competition is of relatively recent origin; the need for a more urgent response becomes crucial as the number of people requiring external support is growing rapidly. Often in situations of great inequality and injustice, accessing the lake should be the strength of the fishermen, and therefore should be studied in the context of the ways in which the fishers construct livelihood systems. A wider understanding of the ways in which fishermen systematically gain access to CPRs, can help inform development interventions, by building on systems already in place that provide benefits to the fishermen. It can also feed into a broader debate, about the ways in which the fishermen can participate, in the extermination of poverty and environmental degradation as actors, rather than as rights of holders.

For fishers, the caste system did extend a certain degree of protection of livelihoods by barring outsiders from entering their occupation (Gadgil and Guha, 1992). However, with the passage of time the influence of the traditional caste system, and their designation of occupational structures has weakened in the Chilika region. This change is significant as I have taken into consideration the influx of outsiders into fishing between the late 1980s and early 1890s, attracted by the success of trawlers. In this context, I have detailed how fishing communities of Chilika perceived their own rights of access to the resources, determined by the ease with which outsiders gain access for the same resources. Fighting against inequality, discrimination and deprivation is a global phenomenon. Collective movements are launched to challenge the unfair exercise of power in society. The current movements in the lake have also emanated from a condition of social unrest experienced by fishers over a period; a situation that has been gathering momentum, in and around Chilika over a period of time. These movements have found resonance with policies governing shrimp gheries and modern commercial ventures, which have threatened the survival of fisher folk. The shrimp ponds in and around Chilika, and the indiscriminate catch of natural shrimp seedlings in the mouth of Chilika are a major threat to the eco-system of the lake. Since they pollute both surface water and ground water, deteriorate the productivity of neighbouring agricultural land, and deplete the estuarine and coastal fishing resources. In addition, they have to contend with siltation and degradation. Entrapped between the commercial interests and the lake's deterioration, the local people are caught between the devil and the deep sea. (Bogaert, 1992). The non-traditional agricultural export (in Stonich's terminology, 1991) in areas around the lake has resulted in capital accumulation, the growing rural impoverishment and the serious problems of environmental degradation. In fact, the fishers traditionally had a positive role in managing the lagoon through indigenous and sustainable fishing systems (Ghosh et al., 2006). The cooperatives further ensured a mechanism of equitable sharing of benefits. The decision of the Supreme Court to change the fishing rights resulted in the loss of perceived rights held by traditional fisher folk. The opinion of the traditional communities was ignored, and the state also ignored the predicaments of fisher folks. The fishers were already hard pressed on account of high handedness of the middlemen who cheated the small scale fisher folk in the process of fish marketing. Indebtedness combined with moneylending added to their misery. According to the first and second NEERI Reports, Suresh Committee Report and UN Report on shrimp culture, while the production and export earnings of the industry grows, the socio-economic losses and environmental degradation affecting the well-being of coastal populations are hardly noticed (Supreme Court Judgment 1996). In this case, I concur with Olson to use collective action as a route of alleviating the misery of Chilika. The core proposition behind this: 'Unless there is coercion or some other special device to make individuals act in their common interest, rational, self-interested individuals will not act to achieve their common or group interests. It argues that the punishments and inducements must be 'selective' so that those who do not contribute can be treated differently from those who do.'

The other point I have highlighted is the mechanization and its impact on the lake ecology which has a direct bearing on the fishery livelihood. As Omvedt, 1993 argues, the environmental destruction has destroyed the conditions of production of peasants, fisher-folk, hunters and gatherers in many parts of Asia in many ways; it has also affected their production itself, and not merely the quality of their daily lives. Climate, food availability and quality, suitability of shelter, and the presence of predators, including humans affect the fish populations. Water quality, turbidity, and the presence of pollutants can also affect fish reproductive success, which directly affects fish populations. Species composition is usually determined by the number of factors including water quality. Fishing has a large impact on fish population and community structure and thus, it must be regulated to maintain a balanced fishery. In short, any human activity that affects water quality and habitat has a potential direct effect on the fish population and overall fishery structure. Thomson (1989) noted that although mechanized and motorized craft allowed larger catches than non-motorized boats, these were often obtained at the expense of the latter. He concluded that motorized fishing operations were unlikely to break even unless additional costs were covered by larger catches, which was quite unlikely in the context of the growing resource depletion.

Further, as I discussed in the last chapter on the environmental issues and natural hazards, these must be dealt in a more careful manner. Over the years, since its exposure to 'Super Cyclone' in

1999, the state of Odisha has progressed effectively towards result-based disaster risk reduction actions; tested and proved in response to Cyclone Phailin (2013) and Cyclone Hudhud (2014). However, when it comes to overall vulnerability and capacity to deal with high impact economic and non-economic losses (exception being protection of life), enormous opportunities of work and improvement await the state. For example, Cyclone Phailin in 2013 caused an economic loss to the tune of Rs. 3,783 million for the fisheries sector in three districts, namely Ganjam, Puri and Khurda.

Hazards have the potential to become disasters except when they encounter strong resilience built by reduced exposure, vulnerabilities and enhanced capacities at all levels. However, the reality is, India as a country and the states still lack basic resilience indicators; thus, the levels of situational complexity and struggle results in high exposure and increased vulnerability among the fishermen. This is deeply linked with their livelihood options, economic advancement, education and knowledge, health, social interactions and many such factors. The International Decade of Natural Disaster Reduction (IDNDR) concluded that "environmental protection, as a component of sustainable development and consistent with poverty alleviation, is imperative in the prevention and mitigation of natural disasters" (ISDR 2002). Planners and practitioners have recognized that the overall objectives of environmental management and mitigation of natural disasters are closely related and both aim to promote sustainable communities. They also recognize that these problems share some common roots. Not many initiatives undertaken for environmental management, poverty alleviation, and disaster risk management aim to manage the physical environment, as well as support sustainable and resilient livelihoods at the community level. In the context of Chilika, the government agencies like the Chilika Development Authority, and the Department of Fisheries and Animal Husbandry, are responsible for overall development and management of fisheries. They need to enhance the scope of visualization to see investment in safety of fishermen as a long-term investment for greater productivity and sustainability of fisheries in the Chilika lagoon.

Abbreviations

1. APL–Above Poverty Line
2. BFDA–Brackish Water Fishery Development Authority
3. BLC–Beach Landing Craft
4. BPL–Below Poverty Line
5. CDA–Chilika Development Authority
6. CPRs–Common Property Resources
7. CAFL–Chilika Aquatic Farms Limited
8. CDS–Catch Documentation Schemes
9. DOF–Department of Fisheries, Government of Odisha
10. DADF–Department of Animal Husbandry Dairying and Fisheries under the Ministry of Agriculture, Government of India
11. FISHFED–Fish Marketing & Processing Federation Ltd.
12. FAO–Food and Agriculture Organization of United Nations
13. FGD–Focused Group Discussion
14. GIS–Geographical Information System
15. IUCN–International Union for Conservation of Nature
16. ICAR–Indian Council for Agricultural Research
17. ICSF–International Collective in Support of Fish workers
18. IUU–Illegal, Unreported and Unregulated
19. IDNDR–International Decade of Natural Disaster Reduction
20. MPEDA–Marine Products Export Development Authority
21. MPEDA–Marine Product Export Development Authority
22. NGOs–Non Government Organizations
23. NEERI–National Environmental Engineering Research Institute
24. NETFISH–Network for Fish Quality Management and Sustainable Fishing

25. OMFRA–Odisha Marine Fisheries Regulation Act
26. OMFRAR–Odisha Marine Fishing Regulation Act and Rules"
27. OSDMA–Odisha State Disaster Management Authority
28. PFCS–Primary Fisher People Cooperation Societies
29. PHC–Petroleum Hydrocarbons Concentration
30. RTE Act–Right to Education Act
31. ROSCA–Rotating Savings and Credit Associations
32. SST–Surface Sea Temperatures
33. SOWFA–The State of World Fisheries and Aquaculture, 2016
34. UN–United Nations
35. UNGA–United Nations General Assembly
36. UNCLOS-United Nations Convention on the Law of the Sea

Index